MACHINE LEARNING MATHEMATICS

Study Deep Learning Through Data Science.

How to Build Artificial Intelligence Through Concepts of Statistics, Algorithms, Analysis, and Data Mining

Samuel Hack

Copyright © 2020 by Samuel Hack - All rights reserved

The book is only for personal use. No part of this publication may be reproduced, distributed, or transmitted in any form or by any means, including photocopying, recording, or other electronic or mechanical methods, without the prior written permission of the publisher, except in the case of brief quotations embodied in critical reviews and certain other noncommercial uses permitted by copyright law.

TABLE OF CONTENTS

INTRODUCTION 7

CHAPTER 1: INTRODUCTION TO MACHINE LEARNING 10

CHAPTER 2: MACHINE LEARNING ALGORITHMS 27

CHAPTER 3: NEURAL NETWORK LEARNING MODELS 44

CHAPTER 4: LEARNING THROUGH UNIFORM CONVERGENCE 68

CHAPTER 5: DATA SCIENCE LIFECYCLE AND TECHNOLOGIES............83

CONCLUSION 106

INTRODUCTION

Congratulations on purchasing *Machine Learning Mathematics*: Study Deep Learning through Data Science. How to Build Artificial Intelligence Through Concepts Of Statistics, Algorithms, Analysis, and Data Mining and thank you for doing so.

The following chapters will discuss the fundamental concepts of machine learning algorithms and the need for machine learning in resolving modern-day business problems. You will find a detailed explanation of the four different types of machine learning algorithms available in the market today along with the importance of machine learning in the first chapter of this book. Representation, evaluation, and optimization make up the three core concepts of machine learning that are explained in detail. You will be introduced to the concept of "Statistical Learning", which is a descriptive statistics-based machine learning framework that can be categorized as supervised or unsupervised.

In chapter 2 of this book titled "Machine Learning Algorithms", you will learn development and application of some of the most popular supervised machine learning algorithms, with explicit details on linear regression, logistic regression, and Naïve Bayes classification algorithms. In chapter 3 titled "Neural Network Learning Models", it will provide you an overarching guide for everything you need to know for successful development of neural network models by learning how to build data pipelines for your machine learning models and then following specific neural network training approaches. The end-to-end process described in this chapter will provide you an overarching view on how to generate your desired machine learning model from scratch with a focus on neural network models. You will also learn the various components and functions at play in the Artificial Neural Network and Perceptron (single neuron-based network) models as well as various applications of these advance and futuristic machine learning models to resolve everyday business problems.

In the 4th chapter of this book titled, "Learning Through Uniform Convergence", we will take a deep dive into the overlap of machine learning with the field of statistics. One of many borrowed statistical concepts used in the development of machine learning models is "Uniform Convergence", which allows the developer to identify the learnability of the problem at hand based on the data sample size using empirical risk minimizers. You will gain a thorough understanding of the concept of "General Setting of Learning" introduced by Vapnik in 1995 and continues to be central to the concept of machine learning development. A statistical explanation of the impact of "Uniform Convergence" on learnability as a prerequisite using finite classes is provided, along with a discussion on potential learnability without "Uniform Convergence".

The final chapter of this book will provide you a holistic overview of various cutting edge data science technologies like data mining and Artificial Intelligence. The most highly recommended lifecycle for structured data science projects is the "Team Data Science Process" (TDSP) is explained in exquisite detail along with various deliverables that need to be generated at every stage. You will also learn how data science is being leveraged by businesses in their decision-making process. The power of artificial intelligence has already started to manifest in our environment and our everyday objects. So you need to learn the difference between Business Intelligence and Data Science technology. This book is filled with real-life examples to help you understand the nitty-gritty of the concepts and names and description of multiple tools that you can further explore and selectively implement in your business to reap the benefits of these cutting-edge technologies.

There are plenty of books on this subject on the market, thanks again for choosing this one! Every effort was made to ensure it is full of as much useful information as possible, please enjoy!

Chapter 1: Introduction to Machine Learning

The notion of Artificial Intelligence Technology is derived from the idea that computers can be engineered to exhibit human-like intelligence and mimic human reasoning and learning capacities, adapting to fresh inputs and performing duties without needing human intervention. The principle of artificial intelligence encompasses machine learning. Machine Learning Technology (ML) refers to the principle of Artificial Intelligence Technology, which focuses mainly on the designed ability of computers to learn explicitly and self-train, identifying information patterns to enhance the underlying algorithm and making autonomous decisions without human involvement. In 1959, the term "machine learning" was coined during his tenure at IBM by the pioneering gaming and artificial intelligence professor, Arthur Samuel.

Machine learning hypothesizes that contemporary computers can be trained using targeted training data sets, which can readily be tailored to create required functionality. Machine learning is guided by a pattern-recognition method where previous interactions and outcomes are recorded and revisited in a way that corresponds to its present position. Because machines are needed to process infinite volumes of data, with fresh data constantly flowing in, they need to be equipped to adapt to the fresh data without being programmed by a person, considering the iterative aspect of machine learning. Machine learning has close relations with the field of Statistics, which is focused on generating predictions using advanced computing tools and technologies. The research of "mathematical optimization" provides the field of machine learning with techniques, theories, and implementation areas. Machine learning is also referred to as "predictive analytics" in its implementation to address business issues. In ML, the "target" is known as "label", while in statistics, it's called "dependent variable". A "variable" in statistics is known as "feature" in ML. And a "feature creation" in ML is known as "transformation" in statistics.

ML technology is also closely related to data mining and optimization. ML and data mining often utilize the same techniques with considerable overlap. ML focuses on generating predictions based on predefined characteristics of the given training data. On the other hand, data mining pertains to the identification of unknown characteristics in a large volume of data. Data mining utilizes many techniques of ML, but with distinct objectives; similarly, machine learning also utilizes techniques of data mining through the "unsupervised learning algorithms" or as a pre-processing phase to enhance the prediction accuracy of the model. The intersection of these two distinct research areas stems from the fundamental assumptions with which they operate. In machine learning, efficiency is generally assessed about the capacity of the model to reproduce known knowledge, while in "knowledge discovery and information mining (KDD)" the main job is to discover new information. An "uninformed or unsupervised" technique, evaluated in terms of known information, will be easily outperformed by other "supervised techniques". On the contrary, "supervised techniques" can not be used in a typical "KDD" task owing to the lack of training data.

Data optimization is another area that machine learning is closely linked with. Various learning issues can be formulated as minimization of certain "loss function" on training data set. "Loss functions" are derived as the difference between the predictions generated by the model being trained and the input data values. The distinction between the two areas stems from the objective of "generalization". Optimization algorithms are designed to decrease the loss of the training data set. The objective of machine learning is to minimize the loss of input data from the real world.

Machine learning has become such a "heated" issue that its definition varies across the world of academia, corporate companies, and the scientific community. Here are some of the commonly accepted definitions from select sources that are extremely known:

- *"Machine learning is based on algorithms that can learn from data without relying on rules-based programming."* – McKinsey.
- *"Machine Learning, at its most basic, is the practice of using algorithms to parse data, learn from it, and then make a determination or prediction about something in the world."* – Nvidia
- *"The field of Machine Learning seeks to answer the question, how can we build computer systems that automatically improve with experience, and what are the fundamental laws that govern all learning processes?"* – Carnegie Mellon University
- *"Machine learning is the science of getting computers to act without being explicitly programmed."* – Stanford University

TYPES OF MACHINE LEARNING

Supervised Machine Learning

The "supervised machine learning" is widely used in predictive big data analysis because they can assess and apply the lessons learned from previous iterations and interactions to new data set. These learning algorithms are capable of labeling all their current events based on the instructions provided to efficiently forecast and predict future events. For example, the machine can be programmed to label its data points as "R" (Run), "N" (Negative) or "P" (Positive). The machine-learning algorithm then labels the input data as programmed and gets the correct output data. The algorithm compares the production of its own with the "expected or correct" output, identifies potential modifications, and resolves errors to make the model more accurate and smarter. By employing methods like "regression", " prediction", "classification", and "boosting of ingredients" to properly train the learning algorithms, any new input data can be fed to the machine as "target" data set to assemble the learning program as desired. This jump-starts the analysis and propels the learning algorithms to create an "inferred feature", which can be used to generate forecasts and predictions based on output values for future events. Financial organizations and banks, for example, depend heavily on machine-learning algorithms to track credit card fraud and foresee the likelihood of a potential customer not making their loan payments on time.

Unsupervised Machine Learning

Companies often find themselves in a situation in which data sources are required to generate a labeled and categorized training data set are unavailable. In these conditions, the use of unsupervised machine learning is ideal. "Unsupervised learning algorithms" are commonly used to describe how the machine can produce "inferred features" to illustrate hidden patterns from an unlabeled and unclassified component in the stack of data. These algorithms can explore the data so that a structure can be defined within the data mass. Although the unsupervised machine learning algorithms are as effective as the supervised learning algorithms in the exploration of input data and drawing insights from it, the unsupervised algorithms are not capable of identifying the correct output. These algorithms can be used to define data outliers; to produce tailor-made product suggestions; to classify text topics using techniques such as "self-organizing maps", "singular value decomposition" and "k-means clustering". Customer identification, for example, customers can be segmented into groups with shared shopping attributes and targeted with similar marketing strategies and campaigns. Consequently, unsupervised learning algorithms are very common in the online marketing industry.

Semi-Supervised Machine Learning

The "semi-supervised machine learning algorithms" are extremely flexible and able to learn from both "labeled" as well as "unlabeled" or raw data. These algorithms are a "hybrid" of supervised and unsupervised ML algorithms. Usually, the training data set consists of predominantly unlabeled data and a tiny portion of labeled data. The use of analytical methods such as the "forecast", "regression" and "classification" in combination with semi-controlled learning algorithms allows the computer to improve its accuracy in learning and training significantly. These algorithms are often used when the production of processed and labeled training data from the raw data set is highly resource-intensive and less cost-effective for the company. Companies are using their systems with semi-supervised learning algorithms to prevent additional personnel and equipment expenses. For example, the application of technology for "facial recognition" requires an enormous quantity of facial data dispersed across multiple input sources. The processing, classification, and labeling of raw data obtained from sources including internet cameras require a lot of resources and thousands of hours to be used as a training data set.

Reinforcement Machine Learning

The "reinforcement machine learning algorithm" learns from its environment and is much more unique than any of the previously discussed machine learning algorithms. Such algorithms perform activities and carefully record the results of each action, either as an error for a failed outcome or a reward for excellent results. The two main characteristics that distinguish the reinforcement learning algorithm are the "trial and error" analysis technique and the "delayed reward" feedback loop. The computer continually analyzes input data using a variety of calculations and transmits a signal of reinforcement for each correct or intended output to eventually optimize the final results. The algorithm creates an easy action and rewards feedback loop for assessing, recording, and learning what activity has been efficient, in that it resulted in right or intended output in a shorter time. The use of such algorithms enables the system to determine optimal conduct automatically and to maximize its effectiveness in a specific context. Therefore, in the disciplines of gaming, robotics, and navigation systems, the reinforcement machine-learning algorithms are heavily utilized.

Importance of Machine Learning

The seemingly unstoppable interest in ML stems from the same variables that have made "data mining" and "Bayesian analysis" more common than ever before. The underlying factors contributing to this popularity are increasing quantities and data varieties, cheaper and more effective computational processing, and inexpensive data storage. To get a sense of how significant machine learning is in our everyday lives, it is simpler to state what part of our cutting edge way of life has not been touched by it. Each aspect of human life is being impacted by the "smart machines" intended to expand human capacities and improve efficiencies. Artificial Intelligence and machine learning technology is the focal precept of the "Fourth Industrial Revolution" that could question our thoughts regarding being "human".

All of these factors imply that models that can analyze larger, more complicated data while delivering highly accurate results in a short time can be produced rapidly and automatically on a much larger scale. Companies can easily identify potential growth opportunities or avoid unknown hazards by constructing desired machine learning models that meet their business requirements. Data runs through the vein of every company. Increasingly, data-driven strategies create a distinction between winning or losing the competition. Machine learning offers the magic of unlocking the importance of business and customer data to lead to actionable measures and decisions that can skyrocket a company's business and market share.

Machine learning has demonstrated over the recent years that many distinct tasks can be automated which were once deemed as activities only people could carry out, such as image recognition, text processing, and gaming. In 2014, Machine Learning and AI professionals believed the board game "Go" would take at least ten years for the machine to defeat its greatest player in the world. But they were proved mistaken by "Google's DeepMind", which showed that machines are capable of learning which moves to take into account even in such a complicated game as "Go". In the world of gaming, machines have seen much more innovations such as "Dota Bot" from the "OpenAI" team. Machine learning is bound to have enormous economic and social impacts on our day to day lives. Complete set of work activities and the entire industrial spectrum could potentially be automated and the labor market will be transformed forever.

"Machine learning is a method of data analysis that automates analytical model building. It is a branch of artificial intelligence based on the idea that systems can learn from data, identify patterns and make decisions with minimal human intervention."

- SAS

Repetitive Learning Automation and Information Revelation

Unlike robotic automation driven by hardware that merely automates manual tasks, machine learning continuously and reliably enables the execution of high quantity, high volume, and computer-oriented tasks. Artificial intelligence machine learning algorithms help to adapt to the changing landscape by enabling a machine or system to learn, to take note of and reduce its previous mistakes. Machine learning algorithm works as a classifier or a forecasting tool to develop unique abilities and to define data pattern and structure. For instance, an algorithm for machine learning has created a model that will teach itself how to play chess and even how to create product suggestions based on consumer activity and behavioral data. This model is so effective because it can easily adjust to any new data set.

Machine learning allows assessment of deeper and wider data sets using neural networks comprising several hidden layers. Just a couple of years ago, a scheme for detecting fraud with countless hidden layers would have been considered a work of imagination. A whole new world is on the horizon with the emergence of big data and unimaginable computer capabilities. The data on the machines is like the gas on the vehicle, more data addition leads to faster and more accurate results. Deep learning models thrive with a wealth of data because they benefit from the information immediately. The machine-learning algorithms have led to incredible accuracy through the «deep neural networks». Increased accuracy is obtained from deep learning, for instance, from the regular and extensive use of smart technology such as "Amazon Alexa" and "Google Search." These "deep neural networks" also boost our healthcare sector. Technologies like image classification and the recognition of objects are now able to detect cancer with the same precision as a heavily qualified radiologist on MRIs.

Artificial intelligence enables the use of big data analytics in combination with the algorithm for machine learning to be enhanced and improved. Data has developed like its currency and can readily become "intellectual property" when algorithms are self-learning. The crude information is comparable to a gold mine in that the more and more you dig, the more you can dig out or extract "gold" or meaningful insights. The use of machine learning algorithms for the data allows faster discovery of the appropriate solutions and can make these solutions more useful. Bear in mind that the finest data will always be the winner, even though everyone uses similar techniques.

"Humans can typically create one or two good models a week; machine learning can create thousands of models a week."

- Thomas Davenport, The Wall Street Journal

CORE CONCEPTS OF MACHINE LEARNING

Today, there are several kinds of ML, but the notion of ML is mainly based on three components "representation", "evaluation", and "optimization". Here are some of the standard concepts that apply to all of them:

Representation

Machine learning models can not directly hear, see, or sense input examples. Data representation is therefore needed to provide a helpful vantage point for the model in the main data attributes. The choice of significant characteristics that best represent data is very essential to train a machine learning model effectively. "Representation" simply refers to the act of "representing" data points to a computer in a language that it understands using a set of classifiers. A classifier may be defined as "a model that inputs a vector of discrete and/or ongoing function values and outputs a single discrete value called "class". To learn from the represented data, a model must have the desired classifier in the training data set or "hypothesis space" that you want the models to be trained on.

The data features used to represent the input are very critical to the machine learning system. Any "classifier" that is external to the hypothesis space cannot be learned by the model. For developing a required machine learning model, data characteristics are so essential that it can easily be the difference between successful and unsuccessful machine learning projects.

A training data set with several independent "features" which are well linked to the "class" can make learning much easier for the machine. On the other side, it may not be easy for the machine to learn from the class with complex functions. This often requires the processing of the raw data so that the desired features for the ML model can be built from it. The method of deriving features from raw data set tends to be the ML project's most time-consuming and laborious component. It is also considered to be the most creative and interesting part of the project where intuition and "trial and error" play just as important a role as the technical requirements. The ML process is not a "one-shot" process of developing and executing a training data set, but an iterative process requiring analysis of the post-execution output, followed by modification of the training data set. Domain specificity is another reason why the training dataset requires comprehensive-time and effort. Training data set to produce predictions based on consumer behavior analysis for an e-commerce platform will be very distinct from the training data set needed to create a self-driving car. Nevertheless, in the industrial sectors, the core machine learning mechanism stays the same. No wonder, there is a lot of research going on to automate the process of feature engineering.

Evaluation

Essentially, in the context of ML "evaluation", is referred to as the method of assessing various hypotheses or models to select one model over another. An "evaluation function" is needed to distinguish between effective classifiers from the vague ones. The evaluation function is also known as the "objective," "utility," or "scoring" function. The machine-learning algorithm has its internal evaluation function that is usually very different from the researchers' external evaluation function used to optimize the classifier. Usually, the evaluation function is described as the first phase of the project before selecting the data representation tool. For example, the self-driving car machine learning model has the feature to identify pedestrians in its vicinity at near-zero, false-negative, and low false-positive rate as an "evaluation function" and the pre-existing condition that needs to be "represented" using applicable data features.

Optimization

The process of searching the hypothesis space of the represented machine learning model to identify the highest-scoring classifier and achieve better evaluation is called "optimization." For algorithms with more than one optimum classifier, selecting the optimization method is very critical in determining the generated classifier and achieving a more effective model of learning. There are a variety of "off-the-shelf optimizers" on the market to kick off new machine learning models before replacing them with custom-designed optimizers.

Statistical Learning Framework

"Statistical learning" is a descriptive statistics-based learning framework that can be categorized as supervised or unsupervised. "Supervised statistical learning" includes constructing a statistical model to predict or estimate output based on single or multiple inputs, on the other hand, "unsupervised statistical learning" involves inputs but no supervisory output, but helps in learning data relationships and structure. One way of understanding statistical learning is to identify the connection between the "predictor" (autonomous variables, attributes) and the "response" (autonomous variable), in order to produce a specific model which is capable of predicting the "response variable (Y)" on the basis of "predictor factors (X)".

"X = f(X) + ε where X = (X1,X2, . . .,Xp)", where "f" is an "unknown function" & "ε" is "random error (reducible & irreducible)".

Here are some fundamental concepts of Statistical Learning:

Prediction and Inference

If there are several inputs "X" easily accessible, but the output "B" production is unknown, "f" is often treated as a black box, provided that it generates accurate predictions for "Y". This is called "prediction". There are circumstances in which we need to understand how "Y" is influenced as "X" changes. We want to estimate "f" in this scenario, but our objective is not simply to generate predictions for "Y". In this situation, we want to establish and better understand the connection between "Y" and "X". Now "f" is not regarded as a black box since we have to understand the underlying process of the system. This is called "inference". In everyday life, various issues can be categorized into the setting of "predictions", the setting of "inferences", or a "hybrid" of the two.

Parametric and Non-parametric Techniques

The "parametric technique" can be defined as an evaluation of "f" by calculating the set parameters (finite summary of the data) while establishing an assumption about the functional form of "f". The mathematical equation of this technique is "f(X) = β0 + β1X1 + β2X2 + . . . + βpXp". The "parametric models" tend to have a finite number of parameters which is independent of the size of the data set. This is also known as "model-based learning". For example, "k-Gaussian models" are driven by parametric technique.

On the other hand, "non-parametric technique" generates an estimation of "f" on the basis of its closeness to the data points, without making any assumptions on the functional form of "f". The "non-parametric models" tend to have a varying number of parameters which grown proportionally with the size of the data set. This is also known as "memory-based learning". For example, "kernel density models" are driven by a non-parametric technique.

Predictions Accuracy and Model Interpretability

Some of the many methods used to learn from statistical data are less adaptable and extremely restrictive. When "inference" is the target, the use of easy and comparatively inflexible techniques of statistical learning has significant benefits. On the other hand, if the target is the generation of forecasts and predictions flexible models are preferred.

The performance of the model can be estimated based on its accuracy to predict the occurrence of an event on new input data. A more accurate model is deemed as a more valuable model. Interpretability of the model offers insight into the input-output relationship. An interpreted model can provide insight into the capability of independent features to generate predictions for the dependent attribute. The problem occurs because, at the expense of interpretability, as model accuracy improves so does the complexity of the model.

A more accurate model can offer a business more possibilities, advantages, time, or money. But the model accuracy needs to be optimized for such prediction. The optimization of accuracy extends the complexity of the model even further by introducing additional model parameters (and resources needed to adjust those parameters). It is much easier and quicker to interpret a model with a relatively small number of parameters. An input coefficient and an intercept term are part of a linear regression model. For instance, every single term can be explored to assess how it contributes to the production of the output. Switching to a logistic regression model provides greater authority In the context of the relationships underlying the potential transformation of a function to output, that too should be explored along with the coefficients.

It is relatively easy to understand a decision tree of small size, but a heavily loaded decision tree needs a distinct perspective to understand why the event is predicted to occur. Furthermore, the optimized combination of several models into one prediction tends to have no significant or timely interpretation. Interpretation is deemed ancillary to model accuracy. For example, models designed to separate and classify "spam" emails from "non-spam" emails as well as models designed to evaluate the price of a real estate.

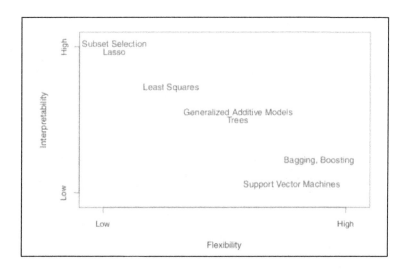

Assessing model accuracy

There is no-one-size-fits-all or jack-of-all-trades technique in the field of statistics, it is impossible for a single technique to dominate across the vast variety of data sets. The most frequently used metric in the "regression" environment is the "mean squared error (MSE)", which can be used for quantification of the extent to which the "predicted answer value" is near to the true answer value for the target observation. For the predicted responses that are extremely close to the true responses, the MSE is computed as small but if the predicted and true responses vary considerably concerning some of the observations then the MSE is computed as large. For example, assume we have clinical data for some patients such as their weight, blood pressure, gender, age, history of family illnesses along with information stating if they are diabetic or not. This patient-related data set can be used to train a statistical technique for predicting diabetes risk based on clinical measures.

$$MSE = \frac{1}{n}\sum_{i=1}^{n}(y_i - \hat{f}(x_i))^2,$$

The most frequently used metric in the "classification" environment is the "confusion matrix". The core characteristic of statistical learning is that with continuous learning the model becomes more flexible and the training errors are reduced, although the test error may not be decreased.

Bias and Variance

In the context of machine learning, bias is defined "the simplified assumptions made by a model to further simplify the learning of the target task". Parametric models are designed with an inherent strong bias, which makes learning much faster and simpler but significantly reduces the overall flexibility of the model for a wider variety of application. "Decision trees", "k-nearest neighbors", and "support vector machines" are categorized as low-bias algorithms for machine learning models. The "high-bias" ML algorithms are "linear regression", "linear discriminant analysis", and "logistic regression".

In the context of machine learning, a variance can be defined "as the amount by which the estimation of the target function will be altered with the use of a different training data set". Non-parametric models" with high flexibility tend to have a high variance score. "Linear regression", "linear discriminant analysis", and "logistic regression" are low-variance ML algorithms. "Decision Trees", "k-Nearest neighbors", and "Support Vector Machines" are high variance ML algorithms.

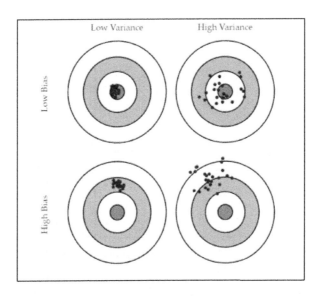

The Trade-Off between Bias and Variance

In the context of statistical learning, "bias" and "variance" are inversely related. For a model exhibiting high bias, the variance score will be reduced significantly and vice versa. There is a compromise that needs to be made between these two factors, which drives the selection of the model and its configuration to resolve the targeted issue by achieving a fine balance between the two. The right level of flexibility is critical to the efficiency and performance of any statistical learning technique in both the "regression" and "classification" environments. The trade-off between "bias" and "variance" of the model and the subsequent "U-shape" in the test error poses a major challenge.

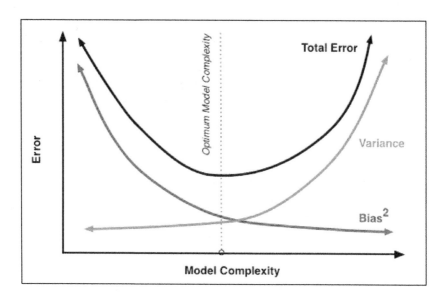

Chapter 2: Machine Learning Algorithms

Machines are now able to learn from and train on their own by using previous computations and underlying algorithms to produce high-quality, easily reproducible decisions and results. Machine learning has been around for a long time now, but the recent developments in machine learning algorithms have made it possible for machines to process and analyze large volumes of data efficiently. This is accomplished by using high speed and frequency automation to apply advanced and complex mathematical calculations to the machines. The sophisticated computing machines of today can rapidly evaluate the ginormous volumes of data and produce faster and more accurate results. Companies that use machine learning algorithms have improved flexibility to adapt the training data set to meet their business requirements and train the machines accordingly. These tailor-made machine learning algorithms allow businesses to identify potential hazards and growth opportunities. Typically, in collaboration with artificial intelligence technology and cognitive technologies, machine learning algorithms are used to produce computers that are highly effective and extremely efficient in processing huge amounts of information or big data and to produce highly accurate results.

Hundreds and thousands of machine learning algorithms have already been generated as this research field continues to proliferate. Here are some of the most commonly used algorithms categorized on the basis of its type of machine learning:

To refresh your memory, "supervised learning" is driven by the data scientists who provide guidance for teaching the algorithm of what conclusions it must make, using predefined training data set.

"Supervised learning" requires information on all possible outputs of the algorithm and the training data set that has already been labeled with expected or correct results.

Let's look at the two of the most renowned supervised learning algorithms used to develop machine learning models, in detail!

REGRESSION

The "regression" techniques fall under the category of supervised machine learning. They help predict or describe a particular numerical value based on the set of prior information, such as anticipating the cost of a property based on previous cost information for similar characteristics. Regression techniques vary from simple (such as "linear regression") to complex (such as "regular linear regression", "polynomial regression", "decision trees", "random forest regression" and "neural networks", among others).

The simplest method of all is "linear regression", where the line's "mathematical equation ($y = m*x+b$) is used to model the data collection". Multiple "data pairs (x, y)" can train a "linear regression" model by calculating the position and slope of a line that can decrease the total distance between the data points and the line. In other words, the calculation of the "slope (m)" and "y-intercept (b)" is used for a line that produces the highest approximation for data observations. The data relationships can be modeled with the use of "linear predictor functions", where unidentified model variables can be estimated from the data. These systems are referred to as "linear models". Traditionally, if values of the "explanatory variables" or "predictors" are known, the conditional mean of the response would be used as the "affinity function" of those values. The use of "conditional media" and other measures in linear models is very rare. Similar to every other form of "regression analysis", the "linear regression" also operates on the "conditional probability distribution" of the responses instead of the joint probability distribution of the variables obtained with the multivariate analysis.

The most rigorously researched form of regression analysis with wide applicability has been "linear regression". This emanates from the fact that models that rely linearly on their unidentified parameters are easy to work with compared to the models that are non-linearly related to their parameters. As the statistical characteristics of the resulting predictors can be easily determined with a linear distribution. There are many useful applications of "linear regression", which can be categorized into one of the following:

If the objective is to generate forecasts and predictions or to reduce errors, the predictive model can be matched to an identified dataset and explanatory variables with the use of a linear regression algorithm. Once the model has been developed, any new input data without a response can be easily predicted by the fitted model.

If the objective is to understand variations in the response variables that may be ascribed to variations in the explanatory variables, "linear regression analysis" could be used to quantify the relationship between the predictors and the response specifically, to assess if certain explanatory variables lack any linear relationship with the response. It can also be used to identify subsets of predictors containing any data redundancies about the response values.

The fitting of most "linear regression models" is accomplished using the "least squares" approach. However, these model can also be fitted by significantly reducing the "lack of fit" in some other standard (just like the "least absolute deviation regression"), or by minimizing a "penalized version of the least square as done in ridge regression (L2-norm penalty) and lasso regression (L1-norm penalty)". By contrast, it is possible to use the "least square" approach to fit machine learning models that are not linear. Therefore, although the terms "least squares" and "linear model" are strongly connected, they are not the same.

"Multiple Linear Regression" tends to be the most common form of "regression" technique used in data science and the majority of statistical tasks. Just like the "linear regression" technique, there will be an output variable "Y" in "multiple linear regression". However, the distinction now is that we're going to have numerous "X" or independent variables generating predictions for "Y".

For instance, a model developed for predicting the cost of housing in Washington DC will be driven by "multiple linear regression" technique. The cost of housing in Washington DC will be the "Y" or dependent variable for the model. "X" or the independent variables for this model will include data points such as vicinity to public transport, schooling district, square footage, and some rooms, which will eventually determine the market price of the housing.

The mathematical equation for this model can be written as below:

"housing_price = $\beta 0 + \beta 1$ sq_foot + $\beta 2$ dist_transport + $\beta 3$ num_rooms"

"Polynomial regression" - Our models developed a straight line in the last two types of "regression" techniques. This straight line is a result of the connection between "X" and "Y" which is "linear" and does not alter the influence "X" has on "Y" as the changing values of "X". Our model will lead in a row with a curve in "polynomial regression".

If we attempted to fit a graph with non-linear features using "linear regression", it would not yield the best fit line for the non-linear features. For instance, the graph on the left is shown in the picture below has the scatter plot depicting an upward trend, but with a curve. A straight line does not operate in this situation. Instead, we will generate a line with a curve to match the curve in our data with a polynomial regression, like the chart on the right shown in the picture below. The equation of a polynomial will appear like the linear equation, the distinction being that one or more of the "X" variables will be linked to some polynomial expression. For instance,

"Y = mX2+b"

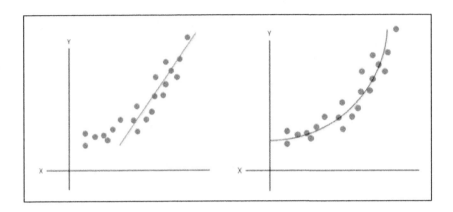

Another significant "regression" technique for data researchers is "Support Vector Regression", which is most frequently used in "case classification". The concept here is to discover a line in space that divides data points into distinct categories. Its also used for regression analysis. It is a form of "binary classification" technique that is not associated with probability.

"Ridge Regression" is a widely used method for analyzing multi-collinear data set. Depending on the features of the data set, using ridge regression correctly can decrease standard errors and significantly improve model accuracy.

Ridge regression can be helpful if your data includes highly correlated independent variables. If you can predict an independent variable with the use of another independent variable, your model will exhibit a high risk of "multi-collinearity". For example, if you use variables that measure the height and weight of a person; these variables in the model are likely to create "multi-collinearity".

Multicollinearity could potentially influence the accuracy of the forecasts and predictions generated by the model. Be mindful of the type of "predictive variables" being utilized in the model to prevent multicollinearity, which could be caused by the type of data you are using, as well as the data collection method. Another reason could be the selection of a small variety of independent variables or the selection of independent variables was very restricted, which resulted in very similar data points.

Multicollinearity can also be induced by generating a highly specific model. Tor that there are more variables than data points in the model. If you have selected to utilize a "linear model" which ended up worsening multicollinearity of the model, then you can attempt to implement a method of "ridge regression".

Ridge regression operates to render the predictions more accurate by permitting a hint of bias into the model. This technique is also referred to as "regularization".

Another technique to enhance the accuracy of the model is by "standardizing" the independent variables. The easiest route is to decrease complexity by changing the values of certain independent variables to null. The approach is not simply to modify these independent variables to null but to implement a structure that rewards values closer to zero. This will trigger the coefficients to decrease, so the model's complexity is also reduced, but the model maintains all of its independent variables. This will offer the model more bias, which is a trade-off for increased accuracy of predictions.

Another technique of reduction is called "LASSO regression". A very complementary technique to the "ridge regression", "lasso regression" promotes the use of simpler and leaner models to generate predictions. In lasso regression, the model reduces the value of coefficients relatively more rigidly. LASSO stands for the "least absolute shrinkage and selection operator". Data on our scatterplot, like the mean or median values of the data are reduced to a more compact level. We use this when the model is experiencing high multicollinearity similar to the "ridge regression" model.

A hybrid of "LASSO" and "ridge regression" methods is known as "ElasticNet Regression". Its primary objective is to further enhance the accuracy of the predictions generated by the "LASSO regression" technique. "ElasticNet Regression" is a confluence of both "LASSO" and "ridge regression" techniques of rewarding smaller coefficient values. All three of these designs are available in the R and Python "Glmnet suite".

"Bayesian regression" models are useful when there is a lack of sufficient data or available data has poor distribution. These regression models are developed based on probability distributions rather than data points, meaning the resulting chart will appear as a bell curve depicting the variance with the most frequently occurring values in the center of the curve. The dependent variable "Y" in "Bayesian regression" is not valuation but a probability. Instead of predicting a value, we try to estimate the probability of an occurrence. This is regarded as "frequentist statistics", and this sort of statistics is built on the "Bayes theorem". "Frequentist statistics" hypothesize if an event is going to occur and the probability of it occurring again in the future.

"Conditional probability" is integral to the concept of "frequentist statistics". Conditional probability pertains to the events whose results are dependent on one another. Events can also be conditional, which means the preceding event can potentially alter the probability of the next event. Assume you have a box of M&Ms and you want to understand the probability of withdrawing distinct colors of the M&Ms from the bag. If you have a set of 3 yellow M&Ms and 3 blue M&Ms, and on your first draw, you get a blue M&M, then with your next draw from the box, the probability of taking out a blue M&M will be lower than the first draw. This is a classic example of "conditional probability". On the other hand, an independent event is flipping of a coin, meaning the preceding coin flip doesn't alter the probability of the next flip of the coin. Therefore, a coin flip is not an example of "conditional probability".

CLASSIFICATION

The "classification algorithm" in machine learning and statistics can be defined as the algorithm used to define the set of categories (sub-populations) that the new input data can be grouped under, based on the training dataset which is composed of related data whose category has already been identified. For instance, all incoming emails can be grouped under the "spam" or "non-spam" category based on predefined rules. Similarly, a patient diagnosis can be categorized based on patient's observed attributes such as gender, blood group, prominent symptoms, family history for genetic diseases. Classification" can be considered as a type of pattern recognition technology. These individual hypotheses can be analyzed into a set of properties that can be easily quantified, referred to as "explanatory variables or features". These can be classified as "categorical", for example different types of blood groups like 'A+', 'O-'; or "ordinal" for example, different types of sizes like large, small; or "integer values", for example, number of times a specific word is repeated in a text; or "real values", for example, height and weight measurement. Certain classifiers operate by drawing a comparison with its prior observations using a "similarity or distance function".

Any machine learning algorithm that is capable of implementing classification, particularly in the context of model implementation, is called "classifier". Very often the term "classifier" is used in the context of the mathematical function, which is being implemented by a "classification algorithm" and can map new input to the appropriate category. In the field of statistics, the classification of data is often carried out with "logistic regression", wherein the characteristics of the observations are referred to as "explanatory variables" or "independent variables" or "regressors" and the categories used to generate predictions are known as "outcomes". These "outcomes" are regarded as the probable values of the dependent variable. In the context of machine learning, "observations are often referred to as instances, the explanatory variables are referred to like features (grouped into a feature vector) and the possible categories to be predicted are referred to as classes".

The "Logistic regression" technique is "borrowed" by ML technology from the world of statistical analysis. "Logistic regression" is regarded to be the simplest algorithm for classification, even though the term sounds like a technique of "regression", that is not the case. "Logistic regression" produces estimates based on single or multiple input values for the likelihood of an event occurring. For example, a "logistic regression" will use a patient's symptoms, blood glucose level, and family history as inputs to generate the likelihood of the patient to develop diabetes. The model will generate a prediction in the form of probability ranging from '1' to '10' where '10' means full certainty. For the patient, if the projected probability exceeds 5, the prediction would be that they will suffer from diabetes. If the predicted probability is less than 5, it would be predicted that the patient will not develop diabetes. Logistic regression allows a line graph to be created which can represent the "decision boundary".

It is widely used for binary classification tasks that involve two different class values. Logistic regression is so named owing to the fundamental statistical function at the root of this technique called the "logistic function". Statisticians created the "logistic function", also called the "sigmoid function", to define the attributes of population growth in ecosystems which continues to grow rapidly and nearing the maximum carrying capacity of the environment. The logistic function is "an S-shaped curve capable of taking any real-valued integer and mapping it to a value between '0' and '1', but never precisely at those boundaries, where 'e' is the base of the natural log (Euler's number or the EXP)" and the numerical value that you are actually going to transform is called the 'value'.

"1 / (1 + e^-value)"

Here is a graph of figures ranging from "-5 and 5" which has been transformed by the logistic function into a range between 0 and 1.

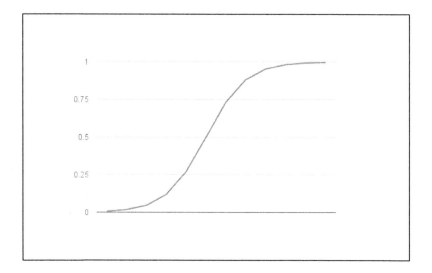

Similar to the "linear regression" technique, "logistic regression" utilizes an equation for data representation.

Input values (X) are grouped linearly to forecast an output value (Y), with the use of weights or coefficient values (presented as the symbol "Beta"). It is mainly different from the "linear regression" because the modeled output value tends to be binary (0 or 1) instead of a range of values.

Below is an example of the "logistic regression" equation, where "the single input value coefficient (X) is represented by 'b1', the "intercept or bias term' is the 'bo', and the "expected result" is 'Y'. Every column in the input data set has a connected coefficient "b" with it, which should be understood by learning the training data set. The actual model representation, which is stored in a file or in the system memory would be "the coefficients in the equation (the beta values)".

$$y=e^{\wedge}(b0 + b1*x)/(1 +e^{\wedge}(b0 + b1*x))$$

The "logistic regression" algorithm's coefficients (the beta values) must be estimated on the basis of the training data. This can be accomplished using another statistical technique called "maximum-likelihood estimation", which is a popular ML algorithm utilized using a multitude of other ML algorithms. "Maximum-likelihood estimation" works by making certain assumptions about the distribution of the input data set.

An ML model that can predict a value nearer to "0" for the "other class" and a value nearer to "1" for the "default class" can be obtained by employing the best coefficients of the model. The underlying assumption for most likelihood of the "logistic regression" technique is that "a search procedure attempts to find values for the coefficients that will reduce the error in the probabilities estimated by the model pertaining to the input data set (e.g. probability of '0' if the input data is not the default class)".

Without going into mathematical details, it is sufficient to state that you will be using "a minimization algorithm to optimize the values of the best coefficients from your training data set". In practice, this can be achieved with the use of an effective "numerical optimization algorithm", for example, the "Quasi-newton" technique).

GENERATING PREDICTIONS USING LOGISTIC REGRESSION

In here, you can simply plug in the measurements into the "logistic regression" equation and calculate the outcome to generate predictions with the "logistic regression" model. Let's take a look at an example to solidify this concept. Let's assume there is a model that is capable of generating predictions if an individual is masculine or woman depending on with fictitious values of their height. If the value of the height for an individual is set as 150 cm, would the individual be predicted as a male or female? Assuming we have already discovered the values of coefficients "b0= -100" and "b1= 0.6". By leveraging the above equation, the probability of male with a height of 150 cm or "P(male|height=150)" can be easily calculated. The function EXP() will be used for "e" because if you log this instance into your spreadsheet, this is what you can use:

"y = e^(b0 + b1*X) / (1 + e^(b0 + b1*X))"

"y = exp(-100 + 0.6*150) / (1 + EXP(-100 + 0.6*X))"

"y = 0.0000453978687"

Or a near "0" probability male is the gender of that specific person.

In theory, probability can simply be used. But since this is a "classification" algorithm and we want a sharp outcome, the probabilities can be tagged on to a binary class value. For instance, the model can predict "0" if "p (male) < 0.51" and predict "1" if "p (male) >= 0.5". Now that you know how predictions can be generated using "logistic regression", you can easily pre-process the training data set to get the most out of this technique. The assumptions made of the distribution and relations within the data set by the "logistic regression" technique are nearly identical to the assumptions made in the "linear regression" technique.

A lot of research has been done to define these hypotheses and to use accurate probabilistic and statistical language. It is recommended to use these as thumb rules or directives and try with various processes for data preparation.

The ultimate goal in "predictive modeling" machine learning initiatives is the generation of highly accurate predictions rather than analysis of the outcomes. Considering everything some assumptions could be broken if the designed model is stable and has high performance.

"Binary Output Variable": This may be evident as we have already discussed it earlier, but "logistic regression" is designed specifically for issues with "binary (two-class) classification". This will generate predictions for the probability of a default class instance that can be tagged into a classification of "0" or "1".

"Remove Noise": Logistic regression does not assume errors in the "output variable ('y')", therefore, the "outliers and potentially misclassified" cases should be removed from the training data set.

"Gaussian Distribution": Logistic regression can be considered as a type of "linear algorithm but with a non-linear transform on the output". A liner connection between the output and input variables is also assumed. Data transforms of the input variables may lead to a more accurate model with a higher capability of revealing the linear relationships of the data set. For instance, to better reveal these relationships, we could utilize "log", "root", "Box-Cox" and other single variable transformations.

"Remove Correlated Inputs": If you have various highly correlated inputs, the model could potentially be "over-fit" similar the "linear regression" technique. To address this issue, you can "calculate the pairwise correlations between all input data points and remove the highly correlated inputs".

"Failure to converge": It is likely for the "expected likelihood estimation" method that is trained on the coefficients to fail to converge. It could occur if the data set contains several highly correlated inputs or there is a very limited data (e.g. loads of "0" in the input data).

"Naïve Bayes classifier algorithm" is another "classification" learning algorithm with a wide variety of applications. It is a method of classification derived from the "Bayes theorem", which assumes predictors are independent of one another. Simply put, a "Naïve Bayes classifier" assumes that "all the features in a class are unrelated to the existence of any other feature in that class". For instance, if input data has an image of a fruit which is green, round, and about 10 inches in diameter, the model can consider the input to be a watermelon. Although these attributes rely on one another or the presence of a specific feature, all of the characteristics contribute independently to the probability that the image of the fruit is that of a watermelon, hence it is referred to as "Naive". "Naïve Bayes model" for large volumes of data sets is relatively simple to construct and extremely effective.

"Naïve Bayes" has reportedly outperformed even the most advanced techniques of classification, along with its simplicity of development. "Bayes theorem" can also provide the means to calculate the posterior probability "P(c|x)" using "P(c), P(x) and P(x)". On the basis of the equation shown in the picture below, where the probability of "c" can be calculated if "x" has already occurred.

"P(c|x)" is the posterior probability of "class (c, target)" provided by the "predictor (x, attributes)". "P(c)" is the class's previous probability. "P(x|c)" is the probability of the class provided by the predictor. "P(x)" is the predictor's prior probability.

$$P(c|x) = \frac{P(x|c)P(c)}{P(x)}$$

where $P(c|x)$ is the Posterior Probability, $P(x|c)$ is the Likelihood, $P(c)$ is the Class Prior Probability, and $P(x)$ is the Predictor Prior Probability.

$$P(c|X) = P(x_1|c) \times P(x_2|c) \times \cdots \times P(x_n|c) \times P(c)$$

Here is an example to better explain the application of the "Bayes Theorem". The picture below represents the data set on the problem of identifying suitable weather days to play golf. The columns depict the weather features of the day and the rows contain individual entries. Considering the first row of the data set, it can be concluded that the weather will be too hot and humid with rain so the day is not suitable to play golf. Now, the primary assumption here is that all these features or predictors are independent of one another. The other assumption being made here is that all the predictors have potentially the same effect on the result. Meaning, if the day was windy it would have some relevance to the decision of playing golf as the rain. In this example, the variable (c) is the class (playing golf) representing the decision if the weather is suitable for golf and variable (x) represents the features or predictors.

	OUTLOOK	TEMPERATURE	HUMIDITY	WINDY	PLAY GOLF
0	Rainy	Hot	High	False	No
1	Rainy	Hot	High	True	No
2	Overcast	Hot	High	False	Yes
3	Sunny	Mild	High	False	Yes
4	Sunny	Cool	Normal	False	Yes
5	Sunny	Cool	Normal	True	No
6	Overcast	Cool	Normal	True	Yes
7	Rainy	Mild	High	False	No
8	Rainy	Cool	Normal	False	Yes
9	Sunny	Mild	Normal	False	Yes
10	Rainy	Mild	Normal	True	Yes
11	Overcast	Mild	High	True	Yes
12	Overcast	Hot	Normal	False	Yes
13	Sunny	Mild	High	True	No

TYPES OF "NAÏVE BAYES CLASSIFIER"

- **"Multinomial Naïve Bayes"** - This is widely used to classify documents, for example, which category does a document belong: beauty, technology, politics, and so on. The frequency of the phrases in the document is considered as the features or predictors of the classifier.
- **"Bernoulli Naive Bayes"** - This is nearly identical to the "Multinomial Naïve Bayes", however, the predictors used here are the "boolean variables". For example, depending on

whether a select phrase occurs in the text or not, the parameters used to predict the class variable can either be a yes or no value.
- **"Gaussian Naive Bayes"** - When the predictors are not distinct and have very similar or continuous values, it can be assumed that these values are obtained from a Gaussian distribution.

APPLICATIONS OF "NAÏVE BAYES"

- **"Real-Time Prediction"**: Naive Bayes is extremely quick in learning from the input data and can be seamlessly used to generate predictions in real-time.
- **"Multi-class Prediction"**: This algorithm is also widely used to generate predictions for multiple classes at the same time. It allows the prediction of the probability of various classes of the target variable.
- **"Text classification / Spam Filtering / Sentiment Analysis"**: The "Naive Bayes classifiers" is heavily used in text classification models owing to its ability to address problems with multiple classes of the target variable and the rule of autonomy. This algorithm has reported higher success rates than any other algorithm. As a consequence, it is commonly used in for identification of spam emails and sentiment analysis by identifying favorable and negative consumer feelings on the social media platforms.
- **"Recommendation System"**: "Naive Bayes Classifier" and "Collaborative Filtering" can be combined to generate a "Recommendation System" that utilizes ML and data mining methods to filter hidden data and generate insight as to whether the customer would prefer a particular item or product.

CHAPTER 3: NEURAL NETWORK LEARNING MODELS

"Artificial Neural Networks" or (ANN) have been developed and designed to mimic the path of communication within the human brain. In the human body, billions of neurons are all interconnected and travel up through the spine and into the brain. They are attached to each other by root-like nodes that pass messages through each neuron one at a time all the way up the chain until it reaches the brain. These systems "learn" to execute jobs by looking at examples, normally without any of the task-specific rules being configured. For instance, they may learn to distinguish pictures that contain dogs using the image recognition technology, by evaluating sample pictures that were manually marked as "dog" or "no dog" and using the outcomes to locate dogs in other pictures. These systems can accomplish this even with no previous understanding of dogs like fur, tails, and dog-like faces. Rather, they are capable of producing identification features automatically from the samples that they are trained on.

An ANN functions as a collection of linked units or nodes called "artificial neurons", that resemble the biological neurons of the human brain. Each link can relay a signal to connected neurons, similar to the synapses in the human brain. An "artificial neuron" receiving a signal can then process it and subsequently transfer it to the connected neurons. When implementing the ANN, the "signal" at a connection will be a real number and the outcome of each neuron will be calculated using certain "non-linear function" of the sum of the inputs. The connections are known as "edges". Generally, the neurons and the "edges" are marked with a value or weight that will be optimized with learning. The weight will increase or decrease the strength of the signal received by the connected neuron. "Concepts" are formed and distributed through the sub-network of shared neurons. Neurons can be set with threshold limits so that a signal will be transmitted only if the accumulated signal exceeds the set threshold. Neurons are usually composed of several layers, which are capable of transforming their inputs uniquely. Signals are passed from the first layer called "input layer" to the final layer called "output layer", sometimes after the layers have been crossed several times.

The initial objective of the ANN model was to resolve problems as accomplished by a human brain. Over time, however, the focus has been directed towards performing select tasks, resulting in a shift from its initial objective. ANNs can be used for various tasks such as "computer vision, speech recognition, machine translation, social media filtering, playing boards, and video games, medical diagnostics, and even painting".

The most common ANN work on a unidirectional flow of information and are called "Feedforward ANN". However, ANN is also capable of the bidirectional and cyclic flow of information to achieve state equilibrium. ANNs learn from past cases by adjusting the connected weights and rely on fewer prior assumptions. This learning could be supervised or non-supervised. With supervised learning, every input pattern will result in the correct ANN output. To reduce the error between the given output and the output generated by ANN, the weights can be varied. For example, reinforced learning, which is a form of "supervised learning", informs the ANN if the generated output is correct instead of providing the correct output directly. On the other hand, unsupervised learning provides multiple input pattern to the ANN, and then the ANN itself explores the relationship between these patterns and learns to categorize them accordingly.

ANNs with a combination of supervised and unsupervised learning are also available.

To solve data-heavy problems where the algorithm or rules are unknown or difficult to comprehend, ANNs are highly useful owing to their data structure and non-linear computations. ANNs are robust to multi-variable data errors and can easily process complex information in parallel. Though, the black-box model of ANN is a major disadvantage, which makes them unsuitable for problems that require a deep understanding and insight into the actual process.

COMPONENTS OF ANNS

- Neurons - ANNs maintained the biological notion of artificial neurons receiving input, combining it with their internal state and threshold value if available, using an "activation function", and generating output using an "output function". Any form of data, including pictures and files, can be used as the initial inputs. The final results obtained can be recognition of an object in a picture. The significant feature of the "activation function" is that as the input values keep changing, it ensures a seamless transition, meaning, a minor change in input will result in a minor change in the output.
- Connections and weights - The ANN comprises of connections that utilize the output from one neuron as an input to an associated neuron. A "weight" that reflects the relative significance of the signal is allocated to each connection. There may be numerous input and output connections for a specific neuron.
- Propagation function - The "propagation function" can calculate the input to a neuron from the output of its predecessors and their connections, in the form of a "weighted sum". A "bias term" may be applied to the

"propagation result". Backpropagation can be defined as a method for adjusting the weights of the connection to adjust for every error encountered, throughout the learning process. The quantity of error is simply split between the connections. Theoretically, "backprop" will calculate the gradient of the "cost function" connected with the weight of the given state. The weight can be updated through the "stochastic gradient descent" or other techniques, including "Extreme Learning Machines", "No-prop" network, "weightless" network, and "non-connectionist" neural network.

HYPERPARAMETER OF ANN

A "hyperparameter" can be defined as a parameter that is set before the actual start of the learning process. The parameter values will be obtained through the process of learning. For example, learning rate, batch size and number of concealed layers. Some "hyperparameter values" could potentially depend on other "hyperparameter values". The size of certain layers, for instance, may rely on the total number of layers. The "learning rate" for each observation indicates the size of the corrective measures that the model requires to compensate for any errors. A higher learning rate reduces the time needed to train the model but results in reduced accuracy. On the other hand, a lower rate of learning increases the time needed to train the model but can result in higher accuracy. Optimizations" like "Quickprop" are mainly directed at accelerating the minimization of errors, while other enhancements are primarily directed at increasing the reliability of the output. Refinements" utilize an "adaptive learning rate" that can increase or decrease as applicable, to avoid oscillation within the network, including the alternation of connection weights, and to help increase the convergence rate. The principle of momentum enables the weighing of the equilibrium between the gradient and the prior alteration so that the weight adjustment will depend on the prior alteration to a certain extent. The gradient is emphasized by the momentum close to "0", while the last change is emphasized by a value close to "1".

NEURAL NETWORK TRAINING WITH DATA PIPELINE

A neural network can be defined as "a function that learns the expected output for a given input from training datasets". Unlike the "Artificial Neural Network", the "Neural Network" features only a single neuron, also called as "perceptron". It is a straightforward and fundamental mechanism which can be implemented with basic math. The primary distinction between traditional programming and a neural network is that computers running on neural network learn from the provided training data set to determine the parameters (weights and prejudice) on their own, without needing any human assistance. Algorithms like "backpropagation" and "gradient descent" may be used to train the parameters. It can be stated that the computer tries to increase or decrease every parameter a bit, in the hope that the optimal combination of Parameters can be found, to minimize the error compared with training data set.

Computer programmers will typically "define a pipeline for data as it flows through their machine learning model". Every stage of the pipeline utilizes the data generated from the previous stage once it has processed the data as needed. The word "pipeline" can be a little misleading as it indicates a unidirectional flow of data, when in reality the machine learning pipelines are "cyclical and iterative", as each stage would be repeated to eventually produce an effective algorithm.

While looking to develop a machine learning model, programmers work in select development environments geared for "Statistics" and 'Machine Learning" such as Python and R among others. These environments enable training and testing of the models, using a single "sandboxed" environment while writing reasonably fewer lines of code. This is excellent for the development of interactive prototypes that can be quickly launched in the market, instead of developing production systems with low latency.

The primary goal of developing a machine learning pipeline is to construct a model with features listed below:

- Should allow for a reduction of system latency.
- Integration but loose coupling with other components of the model, such as data storage systems, reporting functionalities and "Graphical User Interface (GUI)".
- Should allow for horizontal as well as vertical scalability.
- Should be driven by messages, meaning the model should be able to communicate through the transfer of "asynchronous, non-blocking messages".
- Ability to generate effective calculations for management of the data set.
- Should be resilient to system errors and be able to recover with minimal to no supervision, known as breakdown management.
- Should be able to support "batch processing" as well as "real-time" processing of the input data.

Conventionally, data pipelines require "overnight batch processing", which mean gathering the data, transmitting it with an "enterprise message bus" and then processing it to generate pre-calculated outcomes and guidelines for future transactions. While this model has proven to work in certain industrial sectors, in others, and particularly when it comes to machine learning models, "batch processing" doesn't meet the challenge.

The picture below demonstrates a machine learning data pipeline as applied to a real-time business problem in which attributes and projections are dependent on time taken to generate the results. For instance, product recommendation systems used by Amazon, a system to estimate time of arrival used by Lyft, a system to recommend potential new links used by LinkedIn, search engines used by Airbnb, among others.

The swim lane diagram above consists of two explicitly specified components:

1. "Online Model Analytics": In the top swim lane of the picture, the elements of the application required for operation are depicted. It shows where the model is used to make decisions in real-time.

2. "Offline Data Discovery": The bottom swim lane shows the learning element of the model, which is used to analyze historical data and generate the machine learning model using the "batch processing" method.

There are 8 fundamental stages in the creation of a data pipeline, which are shown in the picture below and explained in detail here:

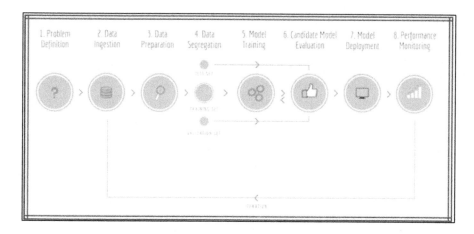

1. Problem Definition

In this stage, the business problem that needs to be resolved using a machine learning model will be identified and documented with all pertinent details.

2. Data Ingestion

The first stage of any machine learning workflow is to channel input data into a database server. The most important thing to remember is that the data is ingested raw and with no modification to enable us to have an invariable record of the original dataset. Data may be supplied from a variety of sources which can be acquired either through request or transmitted from other systems.

"NoSQL document databases" are best suited to store huge amount of defined and labeled as well as unorganized raw data, that are quickly evolving as they do not need to adhere to a predefined scheme. It even provides a "distributed, scalable and replicated data storage".

"Offline"

Data will flow in the "offline" layer to the raw data storage through an "Ingestion Service", which is a "composite orchestration service that is capable of encapsulating the data sourcing and persistence". A repository model is used internally to communicate with a data service that will interact with the data storage in exchange. When you save the data in the database, a unique batch Id will be given to the dataset, which allows for the effective query of the data as well as end-to-end tracking and monitoring of the data.

To be computationally efficient, the ingestion of the data is distributed into two folds.

- The first one is a specific pipeline for every dataset so that each of the datasets can be processed individually and simultaneously.
- The second aspect is that within each pipeline data can be divided to make the best of a variety of server cores, processors and perhaps even the entire server.

Distributing the prepping of data across several vertical and horizontal pipelines will reduce the total time required to perform the tasks.

The "ingestion service" would run at regular intervals based on a predefined schedule (one or more times a day) or upon encountering a trigger. A subject will decouple producers (data source) from processors, which would be the data pipeline for this example, so when the source data is collected, the "producer system" will send a notification to the "broker" and subsequently the "embedded notification service" will respond by inducing ingestion of the data. The "notification service" would also inform the "broker" that the processing of the original dataset was completed with success and now the dataset is being stored in the database.

"Online"

The "Online Ingestion Service" makes up the entrance to the "streaming architecture" of the online layer, as it would decouple and manage the data flow from the source to the processing and storage components by offering consistent, high performance, low latency functionalities. It also works as an enterprise-level "Data Bus". Data would be stored on a long-term "Raw Data Storage", which also serves as a mediating layer to the subsequent online streaming service for further processing in real-time. For instance, such techniques that are utilized in this case may be "Apache Kafka (pub/sub messaging system)" and "Apache Flume (data collection to the long-term database)". A variety of other similar techniques are available and can be selectively applied based on the technology stack of the business.

3. Data Preparation

After the information has been ingested, a centralized pipeline would be produced that can evaluate the condition of the data, meaning it would search for format variations, outliers, patterns, inaccurate, incomplete or distorted information and correct any abnormalities through the process. The "feature engineering process" is also included in this stage. The 3 primary characters of a feature pipeline as depicted in the picture below are: "extraction, transformation, and selection".

Phase	Input	Output
Extract	Raw data	Feature
Transform	Feature	Feature
Select	List<Feature>	List<Feature>

Since this tends to be the most complicated component of any machine learning project, it is essential to introduce appropriate design patterns. In the context of coding, it implies the use of a factory technique to produce features based on certain shared abstract function behavior and a strategy pattern for selecting the correct features at the time of execution can be considered a logical approach. It is important to take into consideration the composition and re-usability of the pipeline while structuring the "feature extractors" and the "transformers".

The selection of functionalities could be attributed to the caller or could be automated. For instance, a "chi-square statistical test" can be applied to classify the impact of each function on the concept label, while discarding the low impact features before starting to train the model. To accomplish this, some "selector APIs" can be identified. In any case, a unique Id must be allocated to each feature set to make sure that the features used as model inputs and for impact scoring are consistent. Overall, it is necessary to assemble a data preparation pipeline into a set of unalterable transformations, which could be readily combined. Now the importance of "testing and high code coverage" will become a critical factor in the success of the model.

4. Data Segregation

The primary goal of the machine learning model is the development of a high accuracy model on the basis of the quality of its forecasts and predictions for information derived from the new input data, which was not part of the training dataset. Therefore, the available labeled dataset will be utilized as a "proxy" for future unknown input data by dividing the data into training and testing datasets. Many approaches are available to split the dataset and some of the most widely used techniques are:

- Using either the default or customized ratio to sequentially divide the dataset into two subsets to ensure that there is no overlap in the sequence in which the data appears from the

source. For example, you could select the first 75% of data to train the model and the consequent 25% of data to test the accuracy of the model.
- Splitting the dataset into training and testing subset using a default or custom ratio with a random seed. For example, you could choose a random 75% of the dataset to train the model and the remaining 25% of the random dataset to test the model.
- Using either of these techniques ("sequential vs. random") and then also mixing the data within each data subset.
- Using a customized injected approach for splitting the data when extensive control over segregation of the data is required.

Technically the data segregation stage is not considered as an independent machine learning pipeline, however, an "API" or tool has to be provided to support this stage. In order to return the required datasets, the next 2 stages ("model training" and "model assessment") must be able to call this "API". As far as the organization of the code is concerned, a "strategy pattern" is required so that the "caller service" can select the appropriate algorithm during execution and the capability to inject the percentage or random seed is required. The "API" must also be prepared to return the information with or without labels, to train and test the model respectively. A warning can be created and passed along with the dataset to secure the "caller service" from defining parameters that could trigger uneven distribution of the data.

5. Model Training

The model pipelines are always "offline", and its schedule will vary from a matter of few hours to just one run per day, based entirely on the complexity of the application. The training can also be initiated by time and event, and not just by the system schedulers.

It includes many libraries of machine learning algorithms such as "linear regression, ARIMA, k-means, decision trees" and many more, which are designed to make provisions for rapid production of new model types as well as making the models interchangeable. Containment is also important for the integration of "third-Party APIs" using the "facade pattern" (at this stage the "Python Jupyter notebook" can also be called).

You have several choices for "parallelization":

- A specialized pipeline for individual models tends to be the easiest method, which means all the models can be operated at the same time.
- Another approach would be to duplicate the training dataset, i.e. the dataset can be divided and each data set will contain a replica of the model. This approach is favored for the models that require all fields of an instance for performing the computations, for example, "LDA", 'MF".
- Another approach can be to parallelize the entire model, meaning the model can be separated and every partition can be responsible for the maintenance of a fraction of the variables. This approach is best suited for linear machine learning models like "Linear Regression", "Support Vector Machine".
- Lastly, a hybrid strategy could also be utilized by leveraging a combination of one or more of the approaches mentioned above.

It is important to implement train the model while taking error tolerance into consideration. The data checkpoints and failures on training partitions must also be taken into account, for example, if every partition fails to owe to some transient problem like timeout, then every partition could be trained again.

TRAINING APPROACHES FOR NEURAL NETWORK

Similar to most of the traditional machine learning models, Neural Networks can be trained using supervised and unsupervised learning algorithms as described below:

Supervised Training

Both inputs and outputs are supplied to the machine as part of the supervised training effort. Then the network will process the inputs and compare the outputs it generated to the expected outputs. Errors will then be propagated back through the model, resulting in the model adjusting the weights that regulate the network. This cycle is repeated time and again with the weights constantly changing. The data set enabling the learning is called the "training set". The same data set is processed several times while the weights of a relationship are constantly improved through the course of training of a network.

Current business network development packages supply resources for monitoring the convergence of an artificial neural network on its capacity to forecast the correct result. These resources enable the training routine to continue for days only until the model reaches the required statistical level or precision. Some networks, however, are incapable of learning. This could be due to the lack of concrete information in the input data from which the expected output is obtained. Networks will also fail to converge if sufficient quantity and quality of the data are not available to confer complete learning. In order to keep a portion of the data set for testing, a sufficient volume of the data set must be available. Most multi-node layered networks can memorize and store large volumes of data. In order to monitor the network to determine whether the system merely retains the training data in a manner that has no significance, supervised learning requires a set of data to be saved and used to evaluate the system once it has been trained.

To avoid insignificant memorization number of the processing elements should be reduced. If a network can not simply resolve the issue, the developer needs to evaluate the inputs and outputs, the number of layers and its elements, the links between these layers, the data transfer and training functionalities, and even the original input weights. These modifications that are needed to develop an effective network comprise the approach in which the "art" of neural networking plays out. Several algorithms are required to provide the iterative feedback needed for weight adjustments, through the course of the training. The most popular technique used is "backward-error propagation", more frequently referred to as "back-propagation". To ensure that the network is not "overtrained", supervised training must incorporate an intuitive and deliberate analysis of the model. An artificial neural network is initially configured with current statistical data trends. Subsequently, it needs to continue to learn other data aspects that could be erroneous from a general point of view. If the model is properly trained and no additional learning is required, weights may be "frozen", if needed. In some models, this completed network is converted into hardware to increase the processing speed of the model. Certain machines do not lock in but continue learning through its use in the production environment.

Unsupervised Training

The network is supplied only with inputs and not the expected results, in "unsupervised or adaptive" training. Then the model must determine its functionality for grouping the input data. This is often called as "self-organization or adaptation". Unsupervised learning is not very well understood at the moment. This adjustment to the surroundings is the pledge that will allow robots to learn continuously on their own as they come across new circumstances and unique settings. Real-world is full of situations wherein there are no training data sets available to resolve a problem. Some of these scenarios include military intervention, which could require new fighting techniques as well as arms and ammunition. Due to this unexpected element of existence and the human desire to be equipped to handle any situation, ongoing study and hope for this discipline continue. However, the vast majority of neural network job is currently carried out in models with supervised learning.

Teuvo Kohonen, an electrical engineer at "Helsinki University of Technology", is one of the pioneering researchers in the field of unsupervised training. He has built a self-organizing network, sometimes referred to as an "auto-associator", which is capable of learning without any knowledge of the expected outcome. It comprises of a single layer containing numerous connections in a network that looks unusual. Weights must be initialized for these connections and the inputs must be normalized. The neurons are organized in the "winner-take-all fashion". Kohonen is conducting ongoing research into networks that are designed differently from the conventional, feed-forward as well as back-propagation methods. Kohonen's research focuses on the organization of neurons into the network of the model.

Neurons within a domain are "topologically organized". Topology is defined as "a branch of mathematics that researches how to map from one space to another without altering its geometric configuration. An instance of a topological organization is the three-dimensional groupings that are common in mammalian brains. Kohonen noted that the absence of topology in designs of neural networks only makes these artificial neural networks a simple abstraction of the real neural networks found in the human brain. Advance self-learning networks could become feasible as this study progresses.

6. Candidate Model Evaluation

The model evaluation stage is also always "offline". By drawing a comparison of the predictions generated with the testing dataset with the actual data values using several key performance indicators and metrics, the "predictive performance" of a model can be measured. To generate a prediction on future input data, the "best" model from the testing subset will be preferred. An evaluator library consisting of some evaluators can be designed to generate accuracy metrics such as "ROC curve" or "PR curve", which can also be stored in a data storage against the model. Once more, the same techniques are applied to make it possible to flexibly combine and switch between evaluators.

The "Model Evaluation Service" will request the testing dataset from the "Data Segregation API" to orchestrate the training and testing of the model. Moreover, the corresponding evaluators will be applied for the model originating from the "Model Candidate repository". The findings of the test will be returned to and saved in the repository. In order to develop the final machine learning model, an incremental procedure, hyper-parameter optimization, as well as regularization methods, would be used. The best model would be deemed as deployable to the production environment and eventually released in the market. The deployment information will be published by the "notification service".

7. Model Deployment

The machine learning model with the highest performance will be marked for deployment for "offline (asynchronous)" and "online (synchronous)" prediction generation. It is recommended to deploy multiple models at the same time to ensure the transfer from obsolete to the current model is made smoothly, this implies that the services must continue to respond to forecast requirements without any lapse, while the new model is being deployed.

Historically, the biggest issue concerning deployment has been pertaining to the coding language required to operate the models have not been the same as the coding language used to build them. It is difficult to operationalize a "Python or R" based model in production using languages such as "C++, C #or Java". This also leads to a major reduction in the performance in terms of speed and accuracy, of the model being deployed. This problem can be dealt with in a few respects as listed below:

- Implementing new language for rewriting the code, for example, "Python to C# translate".
- Creating customized "DSL (Domain Specific Language)" to define the model.
- Creating a "micro-service" that is accessible through "RESTful APIs".
- Implementing an "API first" approach through the course of the deployment.
- Creating containers to store the code independently.
- Adding serial numbers to the model and loading them to "in-memory key-value storage".

In practice, the deployment activities required to implement an actual model are automated through the use of "continuous delivery implementation", which ensures the packaging of the necessary files, validation of the model via a robust test suite as well as the final deployment into a running container. An automated building pipeline can be used to execute the tests, which makes sure that the short, self-containing and stateless unit tests are conducted first. When the model has passed these tests, its quality will be evaluated in larger integrations and by executing regression tests. If both the test phases have been cleared, the model is deemed ready for deployment in the production environment.

Model Scoring

The terms "Model Scoring" and "Model Serving" are being used synonymously throughout the industry. Model Scoring" can be defined as "the process of generating new values with a given model and new input data". Instead of the term "prediction", a generic term "score" is being used to account for the distinct values it may lead to, as listed below:

- List of product recommendations.
- Numerical values, for the "time series models" as well as "regression models".
- A "probability value" that indicates the probability that a new input can be added to a category that already exists in the model.
- An alphabetical value indicating the name of a category that most closely resembles the new input data.
- A "predicted class or outcome" can also be listed as a model score particularly for "classification models".

After the models have been deployed they can be utilized to score based on the feature data supplied by the prior pipelines or provided directly by a "client service". The models must generate predictions with the same accuracy and high performance, in both the online and the offline mode.

"Offline"

The "scoring service" would be optimized in the offline layer for a big volume of data to achieve high performance and generate "fire and forget" predictions. A model can send an "asynchronous request" to initiate its scoring, however, it must wait for the batch scoring process to be completed and gain access to the batch scoring results before the scoring can be started. The "scoring service" will prepare the data, produce the features as well as retrieve additional features from the "Feature Data Store". The results of the scoring will be stored in the "Score Data Store", once the scoring has been completed. The "broker" will be informed of the completion of the scoring by receiving a notification from the service. This event is detected by the model, which will then proceed to collect the scoring results.

"Online"

In the "online" mode, a "client" will send a request to the "Online Scoring Service". The client can potentially request to invoke a specific version of the model, to allow the "Model Router" to inspect the request and subsequently transfer the request to the corresponding model. According to request and in the same way as the offline layer, the "client service" will also prepare the data, produces the features and if needed, fetch additional functions from the "Feature Data Store". When the scoring has been done, the scores will be stored in the "Score Data Store" and then returned via the network to the "client service".

Totally dependent on the use case, results may be supplied asynchronously to the "client", which means the scored will be reported independent of the request using one of the two methods below:

- Push: After the scores have been obtained, they will be pushed to the "client" in the form of a "notification".
- Poll: After the scores have been produced, they will be saved in a "low read-latency database" and the client will poll the database at a regular interval to fetch any existing predictions.

There are a couple of techniques listed below, that can be used to reduce the time taken by the system to deliver the scores, once the request has been received:

- The input features can be saved in a "low-read latency in-memory data store".

- The predictions that have already been computed through an "offline batch-scoring" task can be cached for convenient access as dictated by the use-case, since "offline predictions" may lose their relevance.

8. Performance Monitoring

A very well-defined "performance monitoring solution" is necessary for every machine learning model. For the "model serving clients", some of the data points that you may want to observe include:

- "Model Identifier"
- "Deployment date and time"
- The "number of times" the model was served.
- The "average, min and max" of the time it took to serve the model.
- The "distribution of the features" that were utilized.
- The difference between the "predicted or expected results" and the "actual or observed results".

Throughout the model scoring process, this metadata can be computed and subsequently used to monitor the model performance.

Another "offline pipeline" is the "Performance Monitoring Service", which will be notified whenever a new prediction has been served and then proceed to evaluate the performance while persisting the scoring result and raising any pertinent notifications. The assessment will be carried out by drawing a comparison between the scoring results to the output created by the training set of the data pipeline. To implement fundamental performance monitoring of the model, a variety of methods can be used. Some of the widely used methods include "logging analytics" such as "Kibana", "Grafana" and "Splunk".

A low performing model that is not able to generate predictions at high speed will trigger the scoring results to be produced by the preceding model, to maintain the resiliency of the machine learning solution. A strategy of being incorrect rather than being late is applied, which implies that if the model requires an extended period to time for computing a specific feature then it will be replaced by a preceding model instead of blocking the prediction. Furthermore, the scoring results will be connected to the actual results as they are accessible. This implies continuously measuring the precision of the model and at the same time, any sign of deterioration in the speed of the execution can be handled by returning to the preceding model. In order to connect the distinct versions together, a "chain of responsibility pattern" could be utilized. The monitoring of the performance of the models is an on-going method, considering that a simple prediction modification can cause a model structure to be reorganized. Remember the advantages of machine learning model are defined by its ability to generate predictions and forecasts with high accuracy and speed to contribute to the success of the company.

APPLICATIONS OF NEURAL NETWORK MODELS

Image processing and recognition of character: ANNs have an inherent ability to consume a variety of inputs, which can be processed to derive concealed as well as intricate, non-linear relationships, ANNs play a significant role in the recognition of images and characters. Character recognition such as handwriting adds diverse applicability in the identification of fraudulent monetary transactions and even matters concerning national security. Image recognition is a booming area with extensive applications ranging from facial recognition on the social media platforms like "Instagram" and "Facebook", as well as in medical sciences for detect cancer in patients to satellite image processing for environmental research and agricultural uses. Advancements in the field of ANN has now laid the foundation for "deep neural networks" that serve as the basis for "deep learning" technology.

A variety of groundbreaking and transformative developments in cutting edge technologies like computer vision, voice recognition, and natural language processing has been achieved. ANNs are effective and sophisticated models with a broad variety of utilizations. Some of the real-world applications are:

- "Speech Recognition Network developed by Asahi Chemical"; identification of dementia from the electrode-electroencephalogram study (the leading developer Anderer, noted that the ANN had higher identification accuracy than "Z statistics" and "discriminatory evaluation").
- Generating predictions of potential myocardial infarction in the patient based on their electrocardiogram (ECG). In their research, Baxt and Skora revealed that the diagnostic sensitivity and specificity of the doctors in predicting myocardial infarction was 73.3% and 81.1% respectively, while the diagnostic sensitivity and specificity of the artificial neural network was 96.0% and 96.0% respectively.
- Software for recognition of cursive handwriting utilized by the "Longhand program" of "Lexicus2" executed on current notepads like "NEC Versapad" and "Toshiba Dynapad".
- Optical character recognition (OCR) used by fax machines like "FaxGrabber" developed by "Calera Recognition System" and "Anyfax OCR engine" licensed by "Caere Corporation" to other companies including well known "WinFax Pro" and "FaxMaster".
- "Science Applications International Corporation or (SAIC)" developed the technology to detect bombs in

luggage called "thermal neutron analysis (TNA)", with the use of neural network algorithm.

Forecasting: In day-to-day enterprise decisions, for example, sales forecast, capital distribution between commodities, capacity utilization), economic and monetary policy, finance and inventory markets, forecasting is the tool of choice across the industrial spectrum. For instance, predicting inventory prices is a complicated issue with a host of underlying variables which can be concealed in the depths of big data or readily available. Traditional forecasting models tend to have various restrictions to take these complicated, non-linear associations into consideration. Due to its capacity to model and extract hidden characteristics and interactions, implementation of ANNs in the correct manner can provide a reliable solution to the problem at hand.

ANNs are also free of any restrictions on input and residual distributions, unlike the traditional forecast models. For instance, ongoing progress in this area has resulted in recent advancements in predictive use of "LSTM" and "Recurrent Neural Networks" to generate forecasts from the model. For example, forecasting the weather; foreign exchange systems used by "Citibank London" are driven by neural networks.

Chapter 4: Learning Through Uniform Convergence

The most fundamental issue of statistical learning theory is the issue of characterizing the ability of the model to learn. In the case models driven by "supervised classification and regression" techniques, the learnability can be assumed to be equal to the "uniform convergence" of empirical risk to population risk. Meaning if a problem can be trained on, it can only be learned through minimization of empirical risk of the data. "Uniform convergence in probability", In the context of statistical asymptotic theory and probability theory, is a type of convergence in probability. It implies that within a particular event-family, the empirical frequencies of all occurrences converge to their theoretical probabilities under certain circumstances. "Uniform convergence in probability" as part of the statistical learning theory, is widely applicable to machine learning. Uniform convergence is defined as "a mode of convergence of features stronger than pointwise convergence, in the mathematical field of analysis".

In 1995, Vapnik, published the "General Setting of Learning", as it pertains to the subject of statistical learnability. The General Learning Setting addresses issues of learning. Conventionally, a learning problem can be defined using a "hypothesis class 'H', an instance set 'Z' (with a sigma-algebra), and an objective function (e.g., loss or cost)" i.e. "f: H × Z→ R". This theory can be used "to minimize a population risk functional over some hypothesis class H, where the distribution D of Z is unknown, based on sample $z_1,...,z_m$ drawn from D".

$$\text{"}F(h) = E_{Z \sim D}[f(h;Z)]\text{"}$$

This General Setting encompasses "supervised classification and regression" techniques, some "unsupervised learning algorithms", "density estimation", among others. In supervised learning, "z = (x, y)" is an instance-label pair, "h" is a predictor, and "f (h; (x, y)) = loss(h(x), y)" is the loss function. In terms of statistical learnability, the goal is to minimize "F (h) = EZ~D [f (h;Z)]", within experimental accuracy based on a finite sample only (z1,...zm). The concern, in this case, does not pertain to the problem's computational aspects, meaning whether this approximate minimization can be performed quickly and effectively, but whether this can be achieved statistically based on the sample (z1,...zm) only.

It is common knowledge that a "supervised classification and regression" problem can only be learned when the empirical risks for the whole "h ∈ H" uniformly connect to the population risk. According to the research done by Blumer (1989) and Alon (1997), "if uniform convergence holds, then the empirical risk minimizer (ERM) is consistent, that is, the population risk of the ERM converges to the optimal population risk, and the problem is learnable using the ERM". This suggests that "uniform convergence" of the empirical risks is a required and satisfactory condition for learnability, which can be depicted as an equivocal to a "combinatorial condition" which, when it comes to classification algorithms, it has a finite "VC-dimension" and when in regression algorithms, it has a finite "fat-shattering dimension". The picture below shows a scenario for "supervised classification and regression":

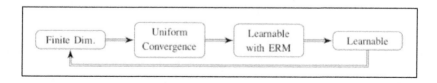

Besides "uniform convergence", "stability's" specific concepts were proposed to be learnability's prerequisite. Inherently, the concepts of "stability" rely on specific "learning laws" or algorithms, and evaluate how sensitive they are to the training data set's fluctuations. It is recognized in specific that ERM stability would be enough for learnability. It is asserted in "Mukherjee et al. (2006)", that stability is also essential to learning. On the basis of the assumption that "uniform convergence is equal to learnability, stability has been shown to characterize learning skill only where uniform convergence characterizes learning skill".

Only in a "supervised classification and regression" environment, was the equivalence of uniform convergence and learning officially established. More broadly, the implications on the "right" in the picture above are valid: "finite fat-shattering dimensions", "uniform convergence", and even "ERM stability", can be deemed suitable for learnability using the "ERM". Concerning the opposite implications, Vapnik has shown that a concept of "non-trivial" or "rigid" learnability associated with the ERM amounts to "uniform convergence of the empirical risks". The concept was intended to banish some of the learning's "trivial" issues that can be learned without uniform convergence. Even with these issues, empirical risk minimization can make learning feasible. Thus, in the "general learning setting" or "supervised classification and regression", an issue would appear to be learnable only when it can be learned by empirical risk minimization.

This framework is not very specific and can sufficiently cover a significant part of the optimization and statistical learning problems that are widely popular, such as:

Stochastic Convex Optimization in Hilbert Spaces: "Let 'Z' be an arbitrary measurable set, let 'H' be a closed, convex, and bounded subset of a Hilbert space, and let 'f(h;z)' be Lipschitz-continuous and convex w.r.t. its first argument. Here, we want to approximately minimize the objective function '$E_{z \sim D}[f(h;z)]$', where the distribution over 'Z' is unknown, based on an empirical sample $z_1,...,z_m$".

Density Estimation: "Let 'Z' be a subset of 'R^n', let 'H' be a set of bounded probability densities on 'Z', and let 'f (h; z) = − log(h(z))'. Here, 'f (·)' is simply the negative log-likelihood of an instance z according to the hypothesis density 'h'. Note that to ensure bounded-ness of 'f (·)', we need to assume that 'h(z)' is lower bounded by a positive constant for all '$z \in Z$'".

K-Means Clustering in Euclidean Space: "Let $Z = R^n$, let 'H' be all subsets of R^n of size k, and let 'f (h; z) = $\min_{c \in h} \|c - z\|^2$'. Here, each h represents a set of 'k centroids', and 'f (·)' measures the Euclidean distance squared between an instance z and its nearest centroid, according to the hypothesis h".

Large Margin Classification in a Reproducing Kernel Hilbert Space (RKHS): "Let '$Z = X \times \{0,1\}$', where 'X' is a bounded subset of an RKHS, let 'H' be another bounded subset of the RKHS, and let 'f (h; (x, y)) = max{0, 1 − y⟨x, h⟩}'. Here, 'f (·)' is the popular hinge loss function, and our goal is to perform margin-based linear classification in the RKHS".

Regression: "Let '$Z = X \times Y$' where 'X' and 'Y' are bounded subsets of 'R^n' and 'R' respectively, let 'H' be a set of bounded functions 'h : $X^n \rightarrow R$', and let '$f(h;(x,y)) = (h(x)-y)^2$'. Here, 'f(·)' is simply the squared loss function".

Binary Classification: "Let '$Z = X \times \{0,1\}$', let 'H' be a set of functions 'h: $X \rightarrow \{0,1\}$', and let 'f (h; (x, y)) = $1_{\{h(x) \neq y\}}$'. Here, 'f (·)' is simply the '0 − 1 loss function', measuring whether the binary hypothesis 'h(·)' misclassified the example (x,y)".

This setting's ultimate goal is a selection of a hypothesis "$h \in H$" based on a finite number of samples with the least amount of potential risk. In general, we are expecting that the sample size will improve the approximation of the risk. It is assumed that "learning guidelines that enable us to choose such hypotheses are consistent". In formal terms, we conclude that "rule A" is consistent with rate "εcons(m)" under distribution "D" if for all "m", where "F* = infh∈H F(h)", the "rate" ε(m) is required to be monotone decreasing with "εcons(m) −→ 0)".

$$"ES \sim Dm\,[F(A(S)) - F*] \leq \varepsilon cons(m)"$$

We can not choose a "D-based" learning rule because "D" is unknown. Rather, we need a "stronger requirement that the rule is consistent with rate εcons(m) under all distributions D over Z". The key definition is as follows:

"A learning problem is learnable, if there exist a learning rule A and a monotonically decreasing m→∞ sequence εcons(m), such that εcons(m) −→ 0, and ∀D, ES~Dm [F(A(S)) − F*] ≤ εcons(m). A learning rule A for which this holds is denoted as a universally consistent learning rule."

The above definition of learnability, which requires a uniform rate of all distributions, is the most appropriate concept to study learnability of a hypothesis class. It is a direct generalization of "agnostic PAC-learnability" to "Vapnik's General Setting of Learning" as studied by Haussler in 1992. A potential path to learning is a minimization of the empirical risk "FS(h)" over a sample "S", defined as

$$"FS(h) = 1/m \sum f(h;z)"$$

Z,z	"Instance domain and a specific instance."
H,h	"Hypothesis class and a specific hypothesis."
f(h,z)	"Loss of hypothesis h on instance z."
B	"$\sup_{h,z} |f(h;z)|$"
D	"Underlying distribution on instance domain Z"
S	"Empirical sample $z_1,...,z_m$, sampled i.i.d. from D"
m	"Size of empirical sample S"
A(S)	"Learning rule A applied to empirical sample S"
$\varepsilon_{cons}(m)$	"Rate of consistency for a learning rule"
F(h)	"Risk of hypothesis h, $E_{z \sim D}[f(h;z)]$"
F*	"$\inf_{h \in H} F(h)$"
$F_S(h)$	"Empirical risk of hypothesis h on sample S, $\frac{1}{m}\sum_{z \in S} f(h;z)$"
$\hat{h}_S$	"An ERM hypothesis, $F_S(\hat{h}_S) = \inf_{h \in H} F_S(h)$ Rate of AERM for a learning rule"
$\varepsilon_{erm}(m)$	
$\varepsilon_{stable}(m)$	"Rate of stability for a learning rule"
$\varepsilon_{gen}(m)$	"Rate of generalization for a learning rule"

The "rule A" is an "Empirical Risk Minimizer" if it can minimize the empirical risk

$$F_S(A(S)) = F_S(\hat{h}_S) = \inf_{h \in H} F_S(h)$$

where "$F_S(\hat{h}_S) = \inf_{h \in H} F_S(h)$" is referred to as the "minimal empirical risk". Given the odds that multiple hypotheses minimize the empirical risk, "$\hat{h}_S$" does not pertain to a certain hypothesis and there could potentially be multiple rules which are all "ERM".

"Rule A" can, therefore, be concluded to be an "AERM (Asymptotic Empirical Risk Minimizer) with rate εerm(m) under distribution D", when:

$$\text{"}E_{S \sim D^m}[F_S(A(S)) - F_S(\hat{h}\,S)] \leq \varepsilon_{erm}(m)\text{"}$$

A learning rule can be considered an "AERM universally" with "rate εerm(m)" if it is an AERM with "rate εerm(m)" under all distributions "D" over "Z". A learning rule can be considered "always AERM" with "rate εerm(m)", if for any "S" sample of "m", size it holds that "FS(A(S))−FS(hˆS) ≤ εerm(m)".

It can be concluded that "rule A" generalizes with rate "εgen(m)" under distribution D if for all m, where A rule "universally generalizes with rate εgen(m) if it generalizes with rate εgen(m) under all distributions D over Z".

$$\text{"}E_{S \sim D^m}[|F(A(S)) - F_S(A(S))|] \leq \varepsilon_{gen}(m)\text{"}$$

IMPACT OF UNIFORM CONVERGENCE ON LEARNABILITY

Uniform convergence is considered applicable to learning problems, "if the empirical risks of hypotheses in the hypothesis class converge to their population risk uniformly, with a distribution-independent rate":

$$\text{"sup}_D E_{S \sim D^m}[\sup_{h \in H} |F(h) - FS(h)|] - m \to \infty \to 0\text{"}$$

It is easy to demonstrate that an issue can be deemed learnable using the "ERM learning rule" if uniform convergence holds.

In 1971, Chervonenkis and Vapnik, demonstrated that "the finiteness of a straightforward combinatorial measure known as the VC dimension indicates uniform convergence, for binary classification issues (where Z = X × {0, 1}, each hypothesis is a mapping from X to {0, 1}, and f (h; (x, y)) = 1{h(x)≠y})". Also, it can be confirmed that in a distribution-independent sense, problems regarding binary classification with infinite "VC-dimension" can be not learned. As a necessary and sufficient condition for learning, this identifies the situation of having finite "VC-dimension", and therefore, uniform convergence.

This characterization is extensible to "regression" techniques as well, namely "regression with squared loss, where h is now a real-valued function, and f (h; (x, y)) = (h(x) − y)2". The property of having a "finite fat-shattering dimension" on all finite scales can substitute for the property of containing "finite VC dimensions", but the basic equivalence still contains, however, a problem can be learned only if there is a uniform convergence. These findings are typically based on sensible reductions made to binary classification. Even though, the "General Learning Setting" observed is not as specific as the classification and regression, including scenarios where it is difficult to reduce the classification to binary classification.

In 1998, Vapnik sought to depict that "in the General Learning Setting, learnability with the ERM learning rule is equivalent to uniform convergence", to bolster the need of uniform convergence in this setting while noting that the result may not hold true to "trivial" situations. Specifically, cases pertaining to "arbitrary learning problem with hypothesis class H and adding H to a single hypothesis h˜ such that f (h˜, z) < inf h∈H f (h, z) for all z ∈ Z", as shown in the picture below. This particular problem of learning can be "trivially" learned using the "ERM learning rule" which always chooses "h˜". Although, "H" can be an arbitrary complex with no prior assumptions and uniform convergence. It must be noted that this is not applicable to binary classification models, where "f (h; (x, y)) = 1{h(x)≠y}", since on any "(x,y)" there will be hypotheses with "f (h;(x,y)) = f (h˜;(x,y))" and therefore, if "H" is highly complex with infinite "VC dimensions then multiple hypotheses will have "0" empirical error on any given training data set.

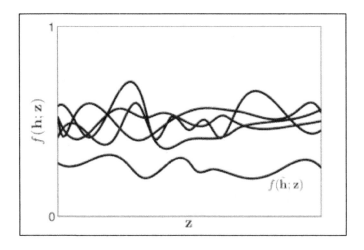

In order to remove such "trivial" scenario, the concept of "strict consistency" was proposed by Vapnik, as an even stronger version of consistency. It is defined with the equation below, where the convergence lies within the probability.

$$"\forall c \in R, \inf_{h:F(h)\geq c} F_S(\mathbf{h}) - m \to \infty \to \inf_{h:F(h)\geq c} F(\mathbf{h})"$$

The thought is that the empirical risk of "ERM" is essential for convergence of the smallest potential risk, even once the "good" hypotheses with less risk than the threshold have been removed. Vapnik succeeded in proving that such "strict consistency" of the ERM has real equivalence to uniform convergence, of the form in probability. The equivalence below is true for every individual distribution and is independent of the universal consistency of the "Empirical Risk Minimizer".

$$"\sup_{h \in H}(F(\mathbf{h}) - F_S(\mathbf{h})) - m \to \infty \to 0"$$

Based on this research study, it can be implied that "up to trivial situations, a uniform convergence property indeed characterizes learnability, at least using the ERM learning rule".

LEARNABILITY WITHOUT UNIFORM CONVERGENCE

A "stochastic convex optimization" or learnability without uniform convergence problem can be considered an exceptional case of the "General Learning Setting" explained above, including additional limitations that "the objective function f(h;z) is Lipschitz-continuous and convex in h for every z, and that H is closed, convex and bounded". The problems where "H" is a subset of a "Hilbert space" will be addressed here. An exceptional scenario is "the familiar linear prediction setting, where z = (x, y) is an instance-label pair, each hypothesis h belongs to a subset H of a Hilbert space, and f (h; x, y) = l(⟨h, φ(x)⟩, y) for some feature mapping φ and a loss function l : R × Y → R, which is convex with respect to its first argument".

The scenario has been successfully established where the stochastic dependence on "h" is linear, similar to the previous example, has been established successfully. When the "domain H" and the "mapping φ" is bounded, there is uniform convergence, meaning that "|F (h) − FS (h)|" is uniformly bounded overall "h ∈ H". This uniform convergence of "FS(h) to F(h)" validates selection of the empirical minimizer "hˆS =arg minh FS (h)", and ensures that the expected value of "F(hˆS)" converges to the optimal value "F∗ = inf h F(h)".

Although dependency on "h" is nonlinear, uniform convergence can still be established with the use of "covering number arguments" provided "H" is a finite dimension. Regrettably, uniform convergence may not take place if we go to infinite-dimensional hypothesis and empirical minimization may not make impart the ability to learn to the algorithm. Remarkably, this does not mean that the problem can be deemed "unlearnable". It can be shown that with regularization mechanisms, even when uniform convergence doesn't exist, a learning algorithm can be developed to solve any "stochastic convex optimization" issue. This mechanism directly relates to the principle of stability. For example, let's look at the "convex stochastic optimization" problem given by the equation in the picture below, where for this example "H" will be the "d-dimensional unit sphere H = h∈Rd : | h| ≤1", "z=(x,α) with α ∈ [0, 1]d" and "x ∈ H" , and "u ∗ v" can be defined as an element-wise product.

$$f^{(3)}(\mathbf{h};(\mathbf{x},\alpha)) = \|\alpha * (\mathbf{h}-\mathbf{x})\| = \sqrt{\sum_i \alpha^2[i](\mathbf{h}[i]-\mathbf{x}[i])^2}$$

Now, considering a series of learning problems, where "d = 2m" for any sample size "m", and establishing that a "convergence rate independent of the dimensionality 'd' cannot be expected". This case can be formalized into infinite dimensions. The learning problem in the equation above can be considered as "that of finding the center of an unknown distribution over x ∈ Rd, where stochastic per-coordinate confidence measures "α[i]" are also available". For now, we will be focusing on the scenario wherein certain coordinates are missing, meaning "α[i] = 0".

By taking into consideration the distribution given below over "(x, α): x = 0" with probability as 1, and "α" is uniform over "{0, 1}d". That is, "α[i]" are independent and identically distributed uniform "Bernoulli". For a random sample "(x1, α1),...,(xm, αm)" if "d > 2m" then that is a result of probability greater than "1 − e−1 > 0.63" and a coordinate "j ∈ 1 . . . d" is present such that all "confidence vectors αi" in the sample are "0" on the coordinate "j", i.e. "αi[j] = 0" for all "i = 1..m". Assume "ej ∈ H" is the "standard basis vector corresponding to this coordinate". Then in the equation shown in the picture below, "FS(3) (·)" represents the empirical risk concerning the function "f(3) (·)".

$$F_S^{(3)}(e_j) = \frac{1}{m}\sum_{i=1}^{m}\|\alpha_i * (e_j - 0)\| = \frac{1}{m}\sum_{i=1}^{m}|\alpha_i[j]| = 0.$$

In another scenario, if "FS(3) (·)" denotes the actual risk for the function "f(3) (·)", the equation shown in the picture below is obtained.

$$F^{(3)}(e_j) = \mathbb{E}_{x,\alpha}\left[\|\alpha * (e_j - 0)\|\right] = \mathbb{E}_{x,\alpha}[|\alpha[j]|] = 1/2$$

Thus, for any sample size "m", a convex "Lipschitz-continuous objective" can be constructed in a dimension that is high enough so as to ensure that with minimum "0.63 probability" over the sample, "suph |F(3) (h)−F(3) (h)| ≥ ½". In addition, since "f (·; ·)" is non-negative, "ej" can be denoted as an "empirical minimizer", even though its expected value "F (3) (ej) = ½" is not at all close to the optimal expected value "minh F(3) (h) = F(3) (0) = 0".

To explain this case with an approach that is not dependent on the sample-size, assume "H is the unit sphere of an infinite-dimensional Hilbert space with orthonormal basis e1, e2,..., where for v ∈ H, we refer to its coordinates v[j] = <v, e j>" with respect to this basis". The "confidences α" serve as a map of every single coordinate to "[0, 1]". This means, an "infinite sequence of reals in [0, 1]". The operation of the product according to the elements, "α * v" is defined on the basis of this mapping and the objective function "f (3) (·)" of the equation (shown in the first picture of this example) can be easily defined in this infinite-dimensional space.

Let us now reconsider the distribution over "z = (x, α)" where "x = 0" and "α" is an infinite independent and identically distributed sequence of "uniform Bernoulli random variables" (that is, a "Bernoulli process with each αi uniform over {0, 1} and independent of all other αj"). It can be implied that for any finite sample there is high likelihood of finding a coordinate "j" with "αi [j] = 0" for all "I", and therefore, an empirical minimizer "FS (3) (ej) = 0" with "F(3) (ej) = 1/2 > 0 = F(3) (0)" can be obtained.

Consequently, it can be observed that the empirical values "FS (3) (h)" are not uniform while converging as expected, and empirical minimization does not guarantee a solution to the learning problem. Furthermore, one could potentially generate a sharper counter-example, wherein the "unique empirical minimizer hˆS" is nowhere close to the optimal expected value. In order to accomplish this, "f(3) (·)" must be augmented with the use of "a small term which ensures its empirical minimizer is unique, and not too close to the origin". Considering the equation below where "ε = 0.01".

$$f^{(4)}(h;(x,\alpha)) = f^{(3)}(h;(x,\alpha)) + \varepsilon \sum 2 - i(h[i] - 1)^2$$

The objective continues to be convex and "(1 + ε)" is still "Lipschitz". In addition, since the added term is strictly convex, the "f(4) (h;z)" will also be strictly convex with respect to "h" and that is the reason for the empirical minimizer being unique.

Considering the same distribution over "z: x = 0" while "α[i]" are independent and identically distributed uniform 0 or 1. The minimizer of "FS (4) (h)" is referred to as the empirical minimizer which is subjected to the constraints "|h| ≤ 1". The good news is that although the identification of the solution for such a constrained optimization problem is complicated, it is not mandatory. It is sufficient to depict that "the optimum of the unconstrained optimization problem h∗UC = arg minFS(4) (h) (with no constraining h ∈ H) has norm |h∗UC| ≥ 1".

It should be noted that "in the unconstrained problem, wherein $a_i[j]$ = 0 for all i = 1...n, only the second term of f (4) depends on h[j] and we have h*UC [j] = 1". As it could happen for certain coordinate "j", it can be concluded that "the solution to the constrained optimization problem lies on the boundary of H , that is $|\hat{h}\ S\ |= 1$", which can be represented by the equation shown in the picture below while "F* ≤ F(0) = ε".

$$F^{(4)}(\hat{h}_S) \geq E_\alpha \left[\sqrt{\sum_i \alpha[i]\hat{h}_S^2[i]} \right] \geq E_\alpha \left[\sum_i \alpha[i]\hat{h}_S^2[i] \right] = \sum_i \hat{h}_S^2[i] E_\alpha [\alpha[i]] = \frac{1}{2} \|\hat{h}_S\|^2 = \frac{1}{2}$$

Chapter 5: Data Science Lifecycle and Technologies

The earliest recorded use of the term data science goes back to 1960 and credited to "Peter Naur", who reportedly used the term data science as a substitute for computer science and eventually introduced the term "datalogy". In 1974, Naur released his book titled "Concise Survey of Computer Methods", with liberal use of the term data science throughout the book. In 1992, the contemporary definition of data science was proposed at "The Second Japanese-French Statistics Symposium", with the acknowledgment of emergence of a new discipline focused primarily on types, dimensions, and structures of data.

The term Data can be defined as "information that is processed and stored by a computer". Our digital world has flooded our realities with data. From a click on a website to our smartphones tracking and recording our location every second of the day, our world is drowning in the data. From the depth of this humongous data, solutions to our problems that we have not even encountered yet could be extracted. This particular process of gathering insights from a measurable set of data using mathematical equations and statistics can be defined as "data science". The role of data scientists tends to be very versatile and is often confused with a computer scientist and a statistician. Essentially, anyone, be it a person or a company, that is willing to dig deep to large volumes of data to gather information can be referred to us data science practitioner. For example, companies like "Amazon" and "Target" keeps a track on and record of in-store and online purchases made by the customers, to provide personalized recommendations on products and services. The social media platforms like "Twitter" and "Instagram", that allow the users to list their current location, is capable of identifying global migration patterns by analyzing the wealth of data that is handed to them by the users themselves.

DATA SCIENCE LIFECYCLE

The most highly recommended lifecycle for structured data science projects is the "Team Data Science Process" (TDSP). This process is widely used for projects that require the deployment of applications based on artificial intelligence and/or machine learning algorithms. It can also be customized for and used in the execution of "exploratory data science" projects as well as "ad hoc analytics" projects. The TDSP lifecycle is designed as an agile and sequential iteration of steps that serve as guidance on the tasks required for the use of predictive models. These predictive models need to be deployed in the production environment of the company, so they can be used in the development of artificial intelligence base applications. The aim of this data science lifecycle is high-speed delivery and completion of data science project toward a defined engagement endpoint. Seamless execution of any data science project requires effective communication of tasks within the team as well as to the stakeholders.

The fundamental components of the "Team Data Science Process" are:

DEFINITION OF A DATA SCIENCE LIFECYCLE

The five major stages of the TDSP lifecycle that outline the interactive steps required for project execution from start to finish are: "Business understanding", "Data acquisition in understanding", "modeling", "deployment" and "customer acceptance". Keep reading for details on this to come shortly!

STANDARDIZED PROJECT STRUCTURE

To enable seamless and easy access to project documents for the team members allowing for quick retrieval of information, use of templates and a shared directory structure goes a long way. All project documents and the project code our store and a "version control system" such as "TFS", "Git" or "Subversion" for improved team collaboration. Business requirements and associated tasks and functionalities are stored in an agile project tracking system like "JIRA", "Rally" and "Azure DevOps" to enable enhanced tracking of code for every single functionality. These tools also help in estimation of resources and costs involved through the project lifecycle. To ensure effective management of each project, information security, and team collaboration, TDSP confers creation of separate storage for each project on the version control system. The adoption of standardized structure for all the projects within an organization, aid in the creation of institutional knowledge library across the organization.

The TDSP lifecycle provides standard templates for all the required documents as well as folder structure at a centralized location. The files containing programming codes for the data exploration and extraction of the functionality can be organized to using the provided a folder structure, which also holds records of model iterations. These templates allow the team members to easily understand the work that has been completed by others as well as for seamless addition of new team members to a given project. The markdown format supports ease of accessibility as well as making edits or updates to the document templates. To make sure the project goal and objectives are well defined and also to ensure the expected quality of the deliverables, these templates provide various checklists with important questions for each project. For example, a "project charter" can be used to document the project scope and the business problem that is being resolved by the project; standardized data reports are used to document the "structure and statistics" of the data.

INFRASTRUCTURE AND RESOURCES FOR DATA SCIENCE PROJECTS

To effectively store infrastructure and manage shared analytics, the TDSP recommends using tools like: "machine learning service", databases, "big data clusters" and cloud-based systems to store data sets. The analytics and storage infrastructure that houses raw as well as processed or cleaned data sets can be cloud-based or on-premises. D analytics and storage infrastructure permits the reproducibility of analysis and prevents duplication and the redundancy of data that can create inconsistency and unwarranted infrastructure costs. Tools are supplied to grant specific permissions to the shared resources and to track their activity which in turn allows secure access to the resources for each member of the team.

TOOLS AND UTILITIES FOR PROJECT EXECUTION

Introduction of any changes to an existing process tends to be rather challenging in most organizations. To encourage and raise the consistency of adoption of these changes several tools can be implemented that are provided by the TDSP. Some of the basic tasks in the data science lifecycle including "data exploration" and "baseline modeling" can be easily automated with the tools provided by TDSP. To allow the hassle-free contribution of shared tools and utilities into the team's "shared code repository", TDSP from provides a well-defined structure. This results in cost savings by allowing other project teams within the organization to reuse and repurpose these shared tools and utilities.

The TDSP lifecycle serves as a standardized template with a well-defined set of artifacts that can be used to garner effective team collaboration and communication across the board. This lifecycle is comprised of a selection of the best practices and structures from "Microsoft" to facilitated successful delivery predictive analytics Solutions and intelligent applications.

Let's look at the details of each of the five stages of the TDSP lifecycle, namely, "Business understanding", "Data acquisition in understanding", "modeling", "deployment" and "customer acceptance".

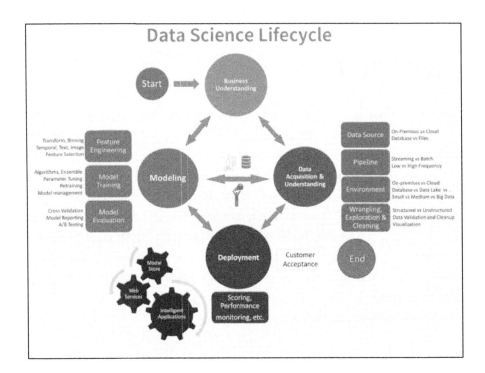

Stage I – Business understanding

The goal of this stage is to gather and drill down on the essential variables that will be used as targets for the model and the metrics associated with these variables will ultimately determine the overall success of the project. Another significant objective of this stage is the identification of required data sources that the company already has or may need to procure. At this stage, the two primary tasks that are required to be accomplished are: "defining objects and identifying data sources".

Deliverables to be created in this stage

- **Charter document** – It is a "living document" that needs to be updated throughout the course of the project, in light of new project discoveries and changing business requirements. A standard template is supplied with the TDSP "project structure definition". It is important to build upon this

document by adding more details throughout the course of the project while keeping the stakeholders promptly updated on all changes made.
- **Data sources** – Within the TDSP "project data report folder", the data sources can be found within the "Raw Data Sources" section of the "Data Definitions Report". The "Raw Data Sources" section also specifies the initial and final locations of the raw data and provide additional details like the "coding scripts" to move up the data to any desired environment.
- **Data dictionaries** – The descriptions of the characteristics and features of the data such as the "data schematics" and available "entity-relationship diagrams", provided by the stakeholders are documented within the Data dictionaries.

Stage II – Data acquisition and understanding

The goal of this stage is the production of high quality processed data set with defined relationships to the model targets and location of the data set in the required analytics environment. At this stage "solution architecture" of the data pipeline must also be developed which will allow regular updates to and scoring of the data. The three primary tasks that must be completed during this stage are: "Data ingestion, Data exploration and Data pipeline set up".

DATA INGESTION

The process required to transfer the data from the source location to the target location should be set up in this phase. The target locations are determined by the environments that will allow you to perform analytical activities like training and predictions.

DATA EXPLORATION

The data set must be scrubbed to remove any discrepancies and errors before it can be used to train the Data models. To check the data quality and gathered information required to process the data before modeling, tools such as data summarization and visualization should be used. Since this process is repeated multiple times, an automated utility called "IDEAR", which is provided by TDSP, can be used for Data visualization and creation of Data summary reports. With the achievement of satisfactory quality of the processed data, the inherent data patterns can be observed. This, in turn, helps in the selection and development of appropriate "predictive model" for the target. Now you must assess if you have the required amount of data to start the modeling process, which is iterative and may require you to identify new data sources to achieve higher relevance and accuracy.

SET UP A DATA PIPELINE

To supplement the iterative process of data modeling, a standard process for scoring new data and refreshing the existing data set must be established by setting up a "data pipeline or workflow". The solution architecture of the data pipeline must be developed by the end of this stage. There are three types of pipelines that can be used on the basis of the business needs and constraints of the existing system: "batch-based", "real-time or streaming" and "hybrid".

DELIVERABLES TO BE CREATED IN THIS STAGE

- **Data quality report** – This report must include "data summary", the relationship between the business requirement and its attributes and variable ranking among other details. The "IDEAR" tool supplied with TDSP it's capable of generating data quality reports on a relational table, CSV file or any other tabular data set.
- **Solution architecture** – A description or a diagram of the data pipeline that is used to score new data and generated predictions, after the model has been built can be referred to as "solution architecture". This diagram can also provide data pipeline needed to "retrain" the model based on new data.
- **Checkpoint decision** – Before that start of the actual model building process project must be reevaluated to determine if the expected value can be achieved by pursuing the project. These are also called as "Go or No-Go" decisions.

Stage III – Modeling

The goal of this stage is to find "optimal data features" for the machine learning model, which is informative enough to predict the target variables accurately and can be deployed in the production environment. The three primary tasks that must be accomplished in this stage are: "feature engineering, model training and the determination of the suitability of the model for the production environment".

Deliverables to be created in this stage

- **Feature sets** – The document containing all the features described in the "feature sets section of the data definition report". It is heavily used by the programmers to write the

required code and develop features based on a description provided by the document.
- **Model report** – This document must contain the details of each model that was evaluated based on a standard template report.
- **Checkpoint decisions** – A decision regarding deployment of the model to the production environment must be made based on the performance of different models.

Stage IV – Deployment

The goal of this stage is to release the solution models to a lower production-like environment such as pre-production environment and user acceptance testing environment before eventually deploying the model in the production environment. The primary task to be accomplished in this stage is "operationalization of the model".

OPERATIONALIZE THE MODEL

Once you have obtained a set of models with expected performance levels, these models can then be operationalized for other applicable applications to use. According to the business requirements, predictions can be made in real-time or on a batch basis. In order to deploy the model, they must be integrated with an open "Application Programming Interface" (API) to allow interaction of the model with all other applications and its components, as needed.

Deliverables to be created in this stage

- A dashboard report using the key performance indicators and metrics to access the health of the system.
- A document or run a book with the details of the deployment plan for the final model.
- A document containing the solution architecture of the final model.

Stage V – Customer Acceptance

The goal of this stage is to ensure that the final solution for the project meets the expectations of the stakeholders and fulfills the business requirements, gathered during the Stage I of the Data science lifecycle. The two primary tasks that must be accomplished in this stage are: "system validation and project hand-off".

Deliverables to be created in this stage

The most important document created during this stage is for the stakeholders and called as "exit report". The document contains all of the available details of the project that are significant to provide an understanding of the operations of the system. TDSP supplies a standardized template for the "exit report", that can be easily customized to cater to specific stakeholder needs.

IMPORTANCE OF DATA SCIENCE

The ability to analyze and closely examine Data trends and patterns using Machine learning algorithms has resulted in the significant application of data science in the cybersecurity space. With the use of data science, companies are not only able to identify the specific network terminal(s) that initiated the cyber attack but are also in a position to predict potential future attacks on their systems and take required measures to prevent the attacks from happening in the first place. Use of "active intrusion detection systems" that are capable of monitoring users and devices on any network of choice and flag any unusual activity, serves as a powerful weapon against hackers and cyber attackers. While the "predictive intrusion detection systems" that are capable of using machine learning algorithms on historical data to detect potential security threats serves as a powerful shield against the cyber predators.

Cyber attacks can result in a loss of priceless data and information resulting in extreme damage to the organization. To secure and protect the data set sophisticated encryption and complex signatures can be used to prevent unauthorized access. Data science can help with the development of such impenetrable protocols and algorithms. By analyzing the trends and patterns of previous cyber attacks on companies across different industrial sectors, Data science can help detect the most frequently targeted data set and even predict potential future cyber attacks. Companies rely heavily on the data generated and authorized by their customers but in the light of increasing cyberattacks, customers are extremely wary of their personal information being compromised and are looking to take their businesses to the companies that are able to assure them of their data security and privacy by implementing advanced data security tools and technologies. This is where data science is becoming the saving grace of the companies by helping them enhance their cybersecurity measures.

Data science has made the use of advanced machine learning algorithms possible which has a wide variety of applicability across multiple industrial domains. For example, the development of self-driving cars that are capable of collecting real-time data using their advanced cameras and sensors to create a map of their surroundings and make decisions of the speed of the vehicle and other driving maneuvers. Companies are always on the prowl to better understand the need of their customers. This is now achievable by gathering the data from existing sources like customer's order history, recently viewed items, gender, age and demographics and applying advanced analytical tools and algorithms over this data to gain valuable insights. With the use of ML algorithms, the system can generate product recommendations for individual customers with higher accuracy. The smart consumer is always looking for the most engaging and enhanced user experience, so the companies can use these analytical tools and algorithms to gain a competitive edge and grow their business.

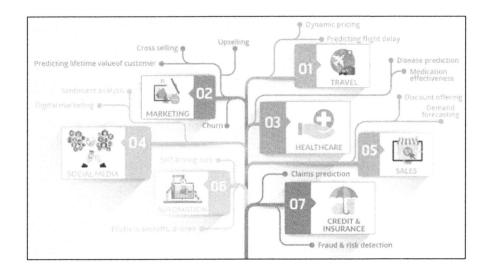

DATA SCIENCE STRATEGIES

Data science is mainly used in decision-making by making precise predictions with the use of "predictive causal analytics", "prescriptive analytics" and machine learning.

Predictive causal analytics – The "predictive causal analytics" can be applied to develop a model which can accurately predict and forecast the likelihood of a particular event occurring in the future. For example, financial institutions use predictive causal analytics-based tools to assess the likelihood of a customer defaulting on their credit card payments, by generating a model that can analyze the payment history of the customer with all of their borrowing institutions.

Prescriptive analytics - The "prescriptive analytics" are widely used in the development of "intelligent tools and applications" that are capable of modifying and learning with dynamic parameters and make their own "decisions". The tool not only predicts the occurrence of a future event but is also capable of providing recommendations on a variety of actions and its resulting outcomes. For example, the self-driving cars gather driving-related data with every driving experience and use it to train themselves to make better driving and maneuvering decisions.

Machine learning to make predictions – To develop models that can determine future trends based on the transactional data acquired by the company, machine learning algorithms are a necessity. This is considered as "supervised machine learning" which we will elaborate on later in this book. For example, fraud detection systems use machine learning algorithms on the historical data on fraudulent purchases to detect if a transaction is fraudulent.

Machine learning for pattern discovery – To be able to develop models that are capable of identifying hidden data patterns but lack required parameters to make future predictions, the "unsupervised machine learning algorithms", such as "Clustering", need to be employed. For example, telecom companies often use the "clustering" technology to expand their network by identifying network tower locations with optimal signal strength in the targeted region.

ARTIFICIAL INTELLIGENCE

Human beings or Homo Sapiens, often tout themselves as being the most superior species to have ever housed the planet Earth, ascribing primarily to their "intelligence". Even the most complex animal behavior is never considered as intelligent, however, the simplest of human behavior is ascribed to intelligence. For example, when a female digger wasp returns with food to her burrow, she deposits the food on the threshold and checks for intruders before carrying her food inside. Sounds like an intelligent wasp right? But an experiment conducted on these wasps, where the scientist displaced the food not too far from the entrance of the burrow while the wasp was inside, revealed that the wasps continued to reiterate the whole procedure every time the food was moved from its original location. This experiment concluded that the inability of the wasp to adapt to the changing circumstances and thereby "intelligence", is noticeably absent in the wasps. So what is this "human intelligence"? Psychologists characterize human intelligence as a composite of multiple abilities such as learning from experiences and being able to adapt accordingly, understanding abstract concepts, reasoning, problem-solving, use of language and perception.

The science of developing human-controlled and operated machinery, such as digital computers or robots, that can mimic human intelligence, adapt to new inputs and perform tasks like humans is called "Artificial Intelligence" (AI). Thanks to Hollywood, most people think of robots coming to life and wreaking havoc on the planet when they hear the words Artificial Intelligence. But that is far from the truth. The core principle of Artificial Intelligence is the ability of the AI-powered machines to rationalize (think like humans) and take actions (mimic human actions) towards fulfilling the targeted goal. Simply put, Artificial Intelligence is the creation of a machine that will think and act like humans. The three paramount goals of Artificial Intelligence are learning, reasoning and perception.

Although the term Artificial Intelligence was coined in 1956, the British pioneer of Computer Sciences, Alan Mathison Turing, performed groundbreaking work in the field of Artificial Intelligence, in the mid-20th century. In 1935, Turing developed an abstract computing machine with a scanner and unlimited memory in the form of symbols. The scanner was capable of moving back and forth through the memory, reading the existing symbols as well as writing further symbols of the memory. A programming instruction would dictate the actions of the scanner and be also stored in the memory. Thus, Turing generated a machine with implicit learning capabilities that could modify and improve its programming. This concept is widely known as the universal "Turing Machine" and serves as a basis for all modern computers. Turing claimed that computers could learn from their own experience and solve problems using a guiding principle known as "heuristic problem solving".

In the 1950s, the early AI research was focused on problem-solving and symbolic methods. By the 1960s, the AI research had a major leap of interest from "The US Department of Defense", who started working towards training computers to mirror human reasoning. In the 1970s, the "Defense Advanced Research Projects Agency" (DARPA) has successfully completed its street mapping projects. It might come to you as a surprise, that DARPA actually produced intelligent personal assistants in 2003, long before the existence of the famous Siri and Alexa. Needless to say, this groundbreaking work in the field of AI has paved the way for automation and reasoning observed in modern-day computers.

Here are the core human traits that we aspire to mimic in the machines:

Knowledge – For machines to be able to act and react like humans, they require an abundance of data and information of the world around us. To be able to implement knowledge engineering AI must have seamless access to data objects, data categories, and data properties as well as the relationship between them that can be managed and stored in the data storages.

Learning – Of all the different forms of learning applicable to AI, the simplest one is "trial and error" method. For example, a chess learning computer program will try all possible moves until the mate-in-one move is found to end the game. This move is then stored by the program to be used the next time it encounters the same position. This relatively easy-to-implement aspect of learning called "rote learning" involves simple memorization of individual items and procedures. The most challenging part of the learning is called "generalization", which involves applying the past experience to the corresponding new scenarios.

Problem Solving – The systematic process to reach a predefined goal or solution by searching through a range of possible actions can be defined as problem-solving. The problem-solving techniques can be customized for a particular problem or used for a wide variety of problems. A general-purpose problem-solving method frequently used in AI is "means-end analysis", which involves a step-by-step deduction of the difference between the current state and final state of the goal. Think about some of the basic functions of a robot, back and forth movement, or picking up stuff that leads to the fulfillment of a goal.

Reasoning – The act of reasoning can be defined as the ability to draw inferences that are appropriate to the given situation. The two forms of reasoning are called "deductive reasoning" and "inductive reasoning". According to "deductive reasoning", if the premise is true then the conclusion is assumed to be true. On the other hand, in the "inductive reasoning", even if the premise is true, the conclusion may or may not be true. Although considerable success has been achieved in programming computers to perform deductive reasoning, the implementation of "true reasoning" remains aloof and one of the biggest challenge facing Artificial Intelligence.

Perception – The process of generating a multidimensional view of an object using various sensory organs can be defined as perception. This creation of awareness of the environment is complicated by several factors, such as the viewing angle, the direction, and intensity of the light and the amount of contrast produced by the object, with the surrounding field. Breakthrough developments have been made in the field of artificial perception and can be easily observed in our daily life with the advent of self-driving cars and robots that can collect empty soda cans while moving through the buildings.

BUSINESS INTELLIGENCE VS. DATA SCIENCE

Data science as you have learned by now is an interdisciplinary approach that applies mathematical algorithms and statistical tools to extract valuable insights from raw data. On the other hand, Business Intelligence (BI) refers to the application of analytical tools and technologies to gain a deeper understanding of the current state of the company as it relates to the company's historical performance. Simply put BI provides intelligence to the company by analyzing their current and historical data whereas data science is much more powerful and capable of analyzing the humongous volume of raw data to make future predictions.

An avalanche of qualitative and quantitative data flowing in from a wide variety of input sources has created a dependency on data science for businesses to make sense of this data and use it to maintain and expand their businesses. The advent of data science as the ultimate decision-making tool goes to show the increasing data dependency for businesses. In some Business intelligence tasks could potentially be automated with the use of data science-driven tools and technologies. The ability to gather insights using these automated tools from anywhere across the world, with the use of the Internet, will only propel the use of "centralized data repositories" for everyday business users.

Business intelligence is traditionally used for "descriptive analysis" and offers retrospective wisdom to the businesses. On the other hand, Data science is much more futuristic and used for "predictive and prescriptive analysis". As data science seeks to answer questions like "Why the event occurred and can it happen again in the future?", Business intelligence focuses on questions like "What happened doing the event and what can be changed to fix it?". It is this fundamental distinction between the "Ws" that are addressed by each of these two fields, that sets them apart.

The niche of business intelligence was once dominated by technology users with computer science expertise, however, Data science is revamping the Business intelligence space by allowing non-technical and core business users to perform analytics and BI activities. Once the data has been operationalized by the data scientists, the tools are easy to use for the mainstream business corridor and can be easily maintained by a support team, without needing any data science expertise. Business intelligence experts are increasingly working hand in hand with the data scientist to develop the best possible Data models and solutions for the businesses.

Unlike Business intelligence that is used to create data reports primarily key performance indicators and metrics dashboards, and provide supporting information for data management strategy, Data science is used to create forecasts and predictions using advanced tools and statistics and provide supplemental information for data governance. A key difference between Data science and business intelligence lies in the range and scale of "built-in machine learning libraries", which empower everyday business user to perform semi-automated or automated data analysis activities. Think of data science as business intelligence on steroids, that is set to turn the business analysis world into a democracy!

Features	Business Intelligence (BI)	Data Science
Data Sources	Structured (Usually SQL, often Data Warehouse)	Both Structured and Unstructured (logs, cloud data, SQL, NoSQL, text)
Approach	Statistics and Visualization	Statistics, Machine Learning, Graph Analysis, Neuro- linguistic Programming (NLP)
Focus	Past and Present	Present and Future
Tools	Pentaho, Microsoft BI, QlikView, R	RapidMiner, BigML, Weka, R

DATA MINING

Data mining can be defined as "the process of exploring and analyzing large volumes of data to gather meaningful patterns and rules". Data mining falls under the umbrella of data science and is heavily used to build artificial intelligence-based machine learning models, for example, search engine algorithms. Although the process of "digging through data" to uncover hidden patterns and predict future events has been around for a long time and referred to as "knowledge discovery in databases", the term "Data mining" was coined as recently as the 1990s.

According to SAS, "unstructured data alone makes up 90% of the digital universe". This avalanche of big data would not essentially guarantee more knowledge. The application of data mining technology allows filtering of all the redundant and unnecessary data noise to garner the understanding of relevant information that can be used in the immediate decision-making process.

Data mining consists of three foundational and highly intertwined disciplines of science, namely, "statistics" (the mathematical study of data relationships), "machine learning algorithms" (algorithms that can be trained with an inherent capability to learn) and "artificial intelligence" (machines that can display human-like intelligence). With the advent of the big data, Data mining technology has been evolved to keep up with the "limitless potential of big data" and relatively cheaper advanced computing abilities. The once considered tedious, labor-intensive, and time-consuming activities have been automated using advance processing speed and power of the modern computing systems.

DATA MINING TRENDS

INCREASED COMPUTING SPEED

With increasing volume and complexity of big data, Data mining tools need more powerful and faster computers to efficiently analyze data. The existing statistical techniques like "clustering" art equipment to process only thousands of input data with a limited number of variables. However, companies are gathering over millions of new data observations with hundreds of variables making the analysis too complicated for the computing system to process. The big data is going to continue to explode, demanding supercomputers that are powerful enough to rapidly and efficiently analyze the growing big data.

LANGUAGE STANDARDIZATION

The data science community is actively looking to standardize a language for the data mining process. This ongoing effort will allow an analyst to conveniently work with a variety of data mining platforms by mastering one standard Data mining language.

SCIENTIFIC MINING

The success of data mining technology in the industrial world has caught the eye of the scientific and academic research community. For example, psychologists are using "association analysis" to capture her and identify human behavioral patterns for research purposes. Economists are using protective analysis algorithms to forecast future market trends by analyzing current market variables.

WEB MINING

Web mining is "the process of discovering hidden data patterns and chains using similar techniques of data mining and applying them directly on the Internet". The 3 main kinds of web mining are: "content mining", "usage mining", and "structure mining". For example, "Amazon" uses web mining to gain an understanding of customer interactions with their website and mobile application to provide more engaging and enhanced user experience to their customers.

DATA MINING TOOLS

Some of the most widely used data mining tools are:

Orange

Orange is "open-source component-based software written in Python". It is most frequently used for basic data mining analysis and offers top-of-the-line data pre-processing features.

RapidMiner

RapidMiner is "open-source component-based software written in Java". It is most frequently used for "predictive analysis" and offers integrated environments for "machine learning", "deep learning" and "text mining".

Mahout

Mahout is an open-source platform primarily used for unsupervised learning process" and developed by "Apache". It is most frequently used to develop "machine learning algorithms for clustering, classification, and collaborative filtering". This software requires advanced knowledge and expertise to be able to leverage the full capabilities of the platform.

MicroStrategy

MicroStrategy is a "business intelligence and data analytics software that can complement all data mining models". This platform offers a variety of drivers and gateways to seamlessly connect with any enterprise resource and analyze complex big data by transforming it into accessible visualizations that can be easily shared across the organization.

CONCLUSION

Thank you for making it through to the end of Machine Learning Mathematics: Study Deep Learning through Data Science. How to Build Artificial Intelligence through Concepts of Statistics, Algorithms, Analysis, and Data Mining. Let's hope it was informative and able to provide you with all of the tools you need to achieve your goals whatever they may be.

The next step is to make the best use of your new-found wisdom on the mathematical or statistical working of the machine learning technology. The fourth industrial revolution is allegedly set to transition the world as we know it, with machines being operated by humans in a limited capacity today into a utopian world out of the science fiction movies, where machines could be indistinguishable from human beings. This transition has been made possible with the power of machine learning. You now have a sound understanding of the statistical learning framework and the crucial role played by uniform convergence and finite classes in determining whether a problem can be resolved using machine learning. To truly capture the essence of machine learning development, expert-level knowledge and understanding of the underlying statistical framework can mean the difference between a successful machine learning model and a failed model that is a time and money sucking machine.

To become a machine learning expert, a solid understanding of the statistical and mathematical concepts of this area is just as important as learning the required programming language. This may seem daunting to most beginners but with this book, we have provided a simplified explanation of the statistical learning framework for ease of understanding. A primary requirement for the development of a winning machine learning algorithm is the quality and generation of the required training data set as well as its learnability by the algorithm. This is the reason we have explained the nuances of training Neural Network in explicit details by building data pipelined from the inception of the project to the implementation and scoring of the model, along with different types of Neural Network training approached. With this knowledge, you are all equipped to design required machine learning algorithms for your business needs. If you are a software developer looking to create that next marvelous application that can learn from the massive amount of open data, competing with the likes of "Amazon Alexa" and "Apple Siri", you have just gotten yourself the head start you were always looking for.

Made in the USA
Coppell, TX
09 December 2020